ACTIVITIES IN ASTRONOMY

Fourth Edition

Darrel B. Hoff
Luther College

Linda J. Kelsey
Center for Naval Analyses

Scott,
May your skies be clear
and filled with stars...
Linda

John S. Neff
The University of Iowa

KENDALL/HUNT PUBLISHING COMPANY
4050 Westmark Drive Dubuque, Iowa 52002

Front Cover Photograph (Credit: Dr. Christopher Burrows, ESA and NASA)

The front cover picture shows the rings associated with Supernova 1987 A taken with the Wide Field Planetary Camera 2 in February, 1994. This image was made following the repair of the Hubble Space Telescope on December 6, 1993. This repair was completed by a four-person repair team, including a 41-year-old, mother of five, Kathy Thornton. The team was led by Story Musgrave who at age 58, is NASA's oldest astronaut. The exquisite detail of this photograph is but one demonstration of the tremendous improvement in the Hubble Telescope following the repair mission.

Supernova 1987 A was the closest supernova to be observed from the earth in nearly 400 years. Exercise 30 in this book has students determining the absolute magnitude of SN 1987A.

Back Cover Photograph (Credit: Robert Sandy, Salem, VA)

The back cover photograph is of the North American Nebula as photographed by Robert Sandy. This demonstrates the type of sky photographs that can be completed with relatively simple equipment. Exercise 6 in this book gives instructions for astronomical photography.

Copyright © 1984, 1992, 1996 by Kendall/Hunt Publishing Company

ISBN 0–7872–0614–8

Printed in the United States of America

10 9 8 7 6 5 4 3 2

DEDICATION

This work is dedicated to Dr. Adrian Docken, Professor Emeritus of Chemistry, and the late Dr. Emil Miller, Professor Emeritus of Physics. They were my advisors and mentors during my undergraduate years at Luther College. To them is owed a most sincere depth of thanks for their efforts on my behalf and on behalf of all Luther students whose undergraduate years were impacted by their tireless service to Luther College.

Darrel Hoff, Ph.D.
Department of Physics
Luther College
Decorah, IA 52101

Contents

Preface and Acknowledgments

This laboratory manual represents an effort to meet the needs of one-semester, introductory astronomy lab courses. This is the fourth edition of the manual. It contains exercises developed over a 30-year period at the University of Iowa, the University of Northern Iowa, the Harvard-Smithsonian Center for Astrophysics, and Luther College. The intent of this manual is to provide exercises that use actual astronomical information so that students will gain a better understanding of the universe. At the same time, students will gain a feeling for the way science is done. Several of these exercises have been shortened or simplified from our two-semester book (*Astronomy: Activities and Experiments*, 2nd Edition), Kelsey, Hoff, and Neff: Dubuque, Iowa: Kendall/Hunt Publishing Company, 1983).

This revised and expanded edition contains 33 exercises with choices for both outdoor observing and classroom work. The book stresses a process theme. What do we mean by a process theme? Too often, beginning students are taught that the end product of science is a *final answer*. We believe that beginning students should also be exposed to the idea that science is an ongoing process and that the problems found in science seldom produce *final answers*. What is most important in science, is the process by which answers are obtained, hence the term, *process approach*. A process approach is generally not communicated to the beginning student in science. The exercises contained herein are designed to stress an open-ended process approach. The exercises are seldom designed to lead to one *correct* answer.

The first section of the book, SKY AND TELESCOPE, contains nine observing labs for clear weather that employ information acquired by naked-eye, binoculars, cameras and small telescopes. One new addition to the book is a ready-to-make star finder. Our experience is that the construction of this simple device gives students a sense of *ownership* and that they are more likely to make sky observations outside of regularly scheduled laboratory periods when they are in possession of their own, self-constructed equipment. We are grateful to Edna DeVore

and the Lawrence Hall of Science for permission to reprint their design of this starfinder.

The book sections, BASIC ASTRONOMY AND THE SOLAR SYSTEM and THE SUN, STARS AND GALAXIES, are designed for classroom use and require little or no special equipment. We encourage students to take their own data and photographs whenever possible. We also include data and photos for use if a lack of equipment or inclement weather prohibits students from obtaining their own. Labs are included containing data on recent developments in astronomy. One lab has students determining the diameters of Pluto and its moon, Charon. Another lab has students calculating the absolute magnitude of SN 1987 A, the closest supernova observed in the last 400 years. In order to strengthen the solar system portion of our book over earlier editions, we have included a lab on determining Mercury's rotation period from radar data and another lab which leads to a determination of the velocity of a comet. Two new labs include one on lunar geology and one on micrometeorite collecting.

A major departure from earlier editions is the provision of pre-labeled graph paper and data sheets following many of the activities. These labeled graph papers and simple answer sheets are provided to assist students who are less-prepared in the use of graphs and completing calculations. For the better-prepared students, these sheets can be ignored and laboratory reports can be prepared following the suggestions for writing laboratory reports found in Appendix 1.

As usual, many persons have contributed to our efforts over the years in producing useful activities in astronomy. In past editions we have credited a number of people for their contributions of ideas, encouragement or materials from which we prepared the laboratory exercises. These people included: Dr. James A. Van Allen, Dr. John D. Fix, Dr. Michael K. Gainer, the late Dr. Charles P. Catalano, Dr. Gary Schmidt, the late Dr. Larry Kelsey, William Lane, Dean Ketelson, Dr. Mark Hodges, Doug Hodges, Mike Haigh, Thomas Wagner, Steve Leiker,

Anne Canaday, Bill Luzader, Joe Toubes and Bill Slinger. The current revision was greatly facilitated by the efforts of Dr. Ardith Hoff. Her fine attention to detail and excellent editing skills have come to be greatly appreciated over the years. We also want to acknowledge the contributions made by the faculty of the University of Florida at Gainsville for use of their observing sheets for Exercises 2 through 5. Allan Morton, Central Arizona College contributed the activity on lunar geology. Mr. Robert Sandy, a masterful photographer of the sky, provided the back cover photo. We add a special thanks to Ms. Kay Weiss of the Kansas Community College, Kansas City for her many helpful comments following the first printing of this 4th edition.

We continue to be most grateful to those administrators under whom we have worked and to numerous people who have given their support and encouragement over the years. We especially thank Dr. James A. Van Allen, former head of the Department of Physics and Astronomy at the University of Iowa, where it all began. Credit also goes to Dr. Wayne Anderson, Head of the Department of Earth Science at the University of Northern Iowa (UNI) where Darrel worked for over 20 years. Darrel's recent work at the Center for Astrophysics at Harvard University, while professionally satisfying, lacked the sincere warmth, openness and trust that he finds here at Luther College. To President H. George Anderson (himself an able amateur astronomer), Dean A. Thomas Kraabel, Associate Dean Mari Heltne, Dr. Dennis Barnaal, Physics Department Head and Dr. Ed Epperly, Education Department Head go my deepest thanks for making a place for me here on the Luther College faculty. To Dr. Richard Kellogg and Mr. Kenneth Larson, I owe a great deal for their helping me get settled in my new position.

Former secretaries, Evelyn Robison (U. of I) and Mary Lou Welch (UNI) are remembered for their past efforts in producing useable manuscripts to submit to the publisher. The current revision was word processed using Macintosh computers.

An instructor's guide has been prepared by Dr. Bruce Palmquist, Central Washington State University and can be obtained by contacting the publisher.

Darrel B. Hoff, Ph.D.
Department of Physics
Luther College
Decorah, IA 52132

Dr. Linda J. Kelsey, Ph.D.
Center for Naval Analyses
4401 Ford Avenue
Post Office Box 16268
Alexandria, VA 22302-0268

John S. Neff, Ph.D.
Department of Physics and Astronomy
University of Iowa
Iowa City, IA 52242

Processes of Science

The common view of science is that it consists of a body of facts; a compendium of ideas and a set of dogmatic concepts. Typically, in beginning science courses, we reinforce this idea by the manner in which we teach the material, and how we test over it. For example, we catalog the number of planets and moons in the solar system, we describe the surface of Mars; we list the features seen on the surface of the sun *etc.* Then we ask you to regurgitate this information. In so doing, we communicate that science is a *product*. In reality, science is only partly product. As practiced by most scientists, it is *process*. This word describes science as a vital, ongoing enterprise. The word conveys the impression that science is dynamic and not static. It says that science is a "doing profession" and that it is in constant change. Unfortunately, this aspect (perhaps the most crucial aspect) is seldom communicated to the beginning student.

What do we really mean by the "process" as part of science? It usually refers to those intellectual skills and practices, which when taken as a group, characterize science. These skills may not be unique only to science, but in aggregate, they are more characteristic of science than any other intellectual discipline.

The Center for Unified Science Education at Ohio State has identified 15 of these processes as follows:

Classifying	Prediction
Inventing Concepts	Hypothesizing
Designing Experiments	Using Numbers
Questioning	Controlling Variables
Identifying Variables	Observing
Interpreting Data	Formulation of Models
Defining Operationally	Using Logic Inferring

In this book, attention will be regularly drawn to this aspect of science. It will be done in three ways. On occasion your instructor will comment on how a certain idea was arrived at, as well as discussing the idea itself. Second, you will be practicing these skills as you do your experiments. Third, you should be on the lookout for examples of the "process" aspect of science as you read your text and go through your course lecture materials.

Darrel B. Hoff
Luther College
Decorah, Iowa

PART 1

Sky and Telescope

1 Visual Astronomy and Star Counts

PURPOSE AND PROCESSES

The purpose of this exercise is to become acquainted with the night sky, to locate some of the more prominent stars and constellations, and to become familiar with some of the literature that is available to assist in naked eye observing. A series of sample star counts will be made in order to estimate the total number of stars on the celestial sphere that are visible to the naked eye. The processes stressed in this exercise include:

> Observing
> Using Numbers
> Inferring

REFERENCES

In addition to the star charts and star finder included in this manual and in your text, there are several other sources of information that are useful in visual observation. These references may be divided into six categories: star finders, periodicals, field books, annuals, atlases, and computer programs. Several are discussed in greater detail in Exercise 9 "Star Charts and Catalogues."

1. **STAR FINDERS:** Since stars exhibit an annual motion (due to the earth's revolution about the sun) and a diurnal motion (due to the earth's rotation on its axis) the locations of stars change from day to day and hour to hour, making the orientation of star charts difficult. Star finders can be adjusted for both the time of day and time of year making the task of locating objects easier. (See Appendix 3 "Equipment Notes.")

2. **PERIODICALS:** The positions relative to the stars of celestial objects such as planets and asteroids change from year to year so that they can not be easily located without consulting a current source of information. Four periodicals which list planetary positions and current celestial events, provide up-to-date information about other subjects of interest to astronomers, and which are commonly used by the beginning student are:

- *Astronomy* [Monthly]
  AstroMedia Corporation
  21027 Crossroads Circle
  Box 1612
  Waukesha, WI 53187-1612

- *Mercury* [Bimonthly]
  Astronomical Society of the Pacific
  390 Ashton Ave
  San Francisco, CA 94112

- *The Planetary Report* [Bimonthly]
  The Planetary Society
  65 North Catalina Ave
  Pasadena, CA 91106

- *Sky and Telescope* [Monthly]
  Sky Publishing Corporation
  P.O. Box 9111
  Belmont, MA 02178-9111

3. **FIELDBOOKS:** These publications provide more comprehensive information about what can be seen in the sky with and without optical aid. They generally provide information about the history of the constellations and the objects in them, nebulae, double stars, colors of stars, and variable stars. Two recommended books are:

- Jay Pasachoff and Donald Menzel, *Field Guide to the Stars and the Planets*, [Boston, MA: Houghton-Mifflin Co., 1993]

- Charles Whitney, *Whitney's Star Finder*, [New York, NY: Alfred A. Knopf, Inc. 1989]

They include additional information about the names of stars, the history of astronomy, information about dwarf and giant stars, meteor showers, periodic comets, and much more.

4. **ANNUALS:** Another useful type of publication is an annual or calendar, which is published yearly, and provides some of the information found in periodicals and field books. They give positions of the planets, phases of the Moon, rising and setting times of the Sun and Moon, locations and properties of the stars, etc. Some commonly used annuals are:

- *The Observer's Handbook,* edited by Roy Bishop, [Royal Canadian Astronomical Society, Toronto, Canada]

- Guy Ottewell, *Astronomical Calendar* [Furman University, Greenville, SC]

Various wall calendars showing historical and celestial events are available from AstroMedia, Sky Publishing, and Optica. Information about celestial events for an upcoming year can also be found in different "farmer" almanacs available at local drug stores and book stores in October or November of the previous year. Addresses are provided in Appendix 3.

5. **ATLASES:** A fifth type of publication is an astronomical atlas. Some commonly used atlases are:

- *Norton's 2000.0 Star Atlas and Reference Book* edited by Ian Ridpath, 18th edition [Essex, England: Longman Scientific and Technical, 1989]

- Wil Tirion, *Sky Atlas 2000.0* [Cambridge, MA: Sky Publishing, Inc., 1981]

- Wil Tirion, *Uranometria 2000.0 Volumes 1 and 2* [Richmond, VA: Willmann-Bell, Inc., 1987]

6. **COMPUTER PROGRAMS:** In recent years computer programs have become available that will calculate the appearance of the sky for a given latitude and longitude for any date and time. Some programs will list the information requested while others show a graphic display of the sky for the time requested. A few even take you to any planet in the solar system to illustrate the motions of its moons. These capabilities are dependent upon the size of the memory and the speed of the microprocessor in the computer. A few of the more popular programs are listed below. Other computer programs are advertised in *Sky & Telescope* or *Astronomy* magazines.

- For IBM
  Dance of the Planets
  A.R.C. Software
  P.O. Box 1974S
  Loveland, CO 80537

- For Macintosh
  Voyager
  Carina Software
  830 Williams St.
  San Leandro, CA 94577

- For the Amiga
  Distant Suns
  Virtual Reality Laboratories
  2341 Ganador Court
  San Luis Obispo, CA 93401

- For the Commodore 64
  Sky Travel
  microillusions
  17408 Chatsworth St.
  Granada Hills, CA 91344

- Various programs for Commodore, Apple II, and IBM are available from:
  Zephyr Services
  1900 Murray Ave. Dept A
  Pittsburgh, PA 15217

INTRODUCTION

1. CONSTELLATIONS

The Greeks recognized the impossibility of attempting to learn much about the heavens without first organizing their information about the vast number of stars in some systematic manner. The geometric arrangement of some stars provided the Greeks with a natural organizational system which we call the *constellations*. These are accidental groupings of stars whose outlines have been given the names of people, objects, and animals they were thought to resemble or chosen to represent. Although the modern astronomer no longer employs these constellations as the ancients did for mythological or astrological purposes, he does use them for quick reference purposes. A number of bright stars are named in the Bayer system making use of the constellation names, and current celestial events are frequently given names of the constellation in which they occur. Novae, for example, are named for the constellation in which they are observed and the year of their occurrence (such as Nova Serpens—1970 or Nova Puppis—1919).

Originally the constellations did not encompass all of the stars, but in 1928 the International Astronomical Union divided the entire celestial sphere into 88 constellations using regular north-south or east-west boundaries so that all stars (and areas of the sky) are now assigned to a constellation.

2. STAR COUNTS

Further observations can be made with the naked eye to study the structure of the universe. We can try to determine the total number of stars visible to the naked eye and visible in telescopes of various sizes, the distribution of stars on the celestial sphere, and the relative number of stars of various magnitudes.

To actually count the stars we can see would be an impossible task. We can simplify the problem by counting the number of stars in several small regions of the sky. To estimate the total number of stars visible, we can calculate the ratio of the area of these regions to the area of the whole celestial sphere. A cardboard tube can be used for observation, and the area of sky being observed can be calculated from the length and radius of the tube.

PROCEDURE

---

1. CONSTELLATIONS

Pull out the two cardboard sheets in Appendix 4. Cut out, fold, and paste **PART 1** of the star wheel holder. Cut out the star-date wheel from **PART 2** and slip it into the holder. The shell should rotate freely, matching dates with times on the holder. By matching the time of night with a specific date, the stars visible in the sky for that date and time are visible in the oval window. This view is for a mid-northern latitude. The center of the opening represents straight-up or zenith. The edge of the opening represents the horizon with the cardinal directions indicated (Figure 1-1).

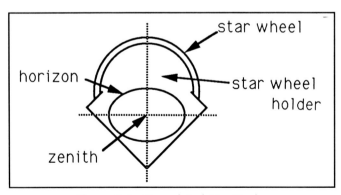

Figure 1-1. Completed Star Finder

Example:

Rotate the star wheel until midnight is pointing at December 10. The constellation Orion, with the three stars in his belt, is about halfway between the southern

horizon and zenith; the brightest star in Auriga is close to zenith (use the SC2 chart in Appendix 4 to identify this star as Capella); and Ursa Major is low in the north-northeast with the handle of the Big Dipper pointing toward the horizon.

As you rotate the star wheel (counter-clockwise), notice that stars are rising from the eastern horizon and setting toward the western horizon. You should also notice that there is a group of constellations in the northern sky that are not rising or setting, they move in circles around the north celestial pole (close to Polaris). These stars that never rise or set are called "circumpolar stars." To study circumpolar stars more comfortably, hold the star finder with "NORTH" down.

You should be able to locate the common circumpolar constellations and the constellations that are associated with each season of the year. Since this exercise is often done in the fall, some general techniques will be discussed to assist in locating the circumpolar constellations and to locate the constellations of fall. Similar techniques can be used for other seasons. Table 1-I lists the constellations and bright stars you should be able to locate each season. Use this star finder in the following problems, matching the date with the time given at the beginning of each section. Use the SC1 and SC2 charts in Appendix 4 to identify specific stars in a given constellation.

(a) Circumpolar Constellations for a Northern Mid-Latitude Observer: It is easier to locate a celestial object if some prominent constellation can be found and used as a reference point. Ursa Major (the Big Dipper) serves as a good reference point. Rotate the star

Table 1-1
Constellations and Objects in Them

Circumpolar Constellations
Ursa Major (Alcor, Mizar, Duhbe, Merak)
Ursa Minor (Polaris)
Cassiopeia (Caph)
Cepheus
Draco (Thuban)

Fall Constellations
Lyra (Vega)
Cygnus (Deneb)
Aquila (Altair)
Delphinus
Sagitta
Pegasus
Capricornus
Aquarius

Winter
Andromeda
Aries
Pisces
Cetus
Perseus (h and χ clusters)
Auriga (Capella)
Taurus (Aldebaran, Pleiades Cluster)
Orion (Betelgeuse, Bellatrix, Rigel)

Spring
Gemini (Pollux, Castor)
Canis Minor (Procyon)
Canis Major (Sirius)
Cancer (Beehive Cluster)
Leo (Denebola, Regulus)
Hydra
Corvus

Summer
Coma Berenices
Bootes (Arcturus)
Libra
Scorpius (Antares)
Sagittarius
Corona Borealis
Serpens
Ophiuchus
Hercules (M 13 Cluster)

wheel so that 9:30 p.m. points at September 15th. Ursa Major lies nearly parallel to and quite close to the northern horizon. The stars at the end of the dipper's cup will be to the east and the cup will open upward. The two stars farthest to the east are called the "pointer stars." They are separated by five degrees of arc (the approximate width of a closed fist held at arm's length). Use these two stars and trace a line upward for a distance of about 30° (or six "fists") to locate the brightest star in Ursa Minor (the Little Dipper). Ursa Minor lies nearly parallel to Ursa Major, and the dippers open toward each other. The brightest star in Ursa Minor is called Polaris and its location is near the north celestial pole.

Locate the pointer stars in Ursa Major, follow them across Polaris to about 30° on the other side of the pole from Ursa Major, and find a large "W" lying on its side. This is the constellation Cassiopeia. Winding its way between the two dippers is an irregular line of stars bearing the name of Draco (the dragon). The fifth circumpolar constellation is composed of an irregular square topped by a triangle of stars, and is called Cepheus. In the autumn Cepheus is located above the pole and west of Cassiopeia with its point directed toward Polaris.

Locate all the stars and constellations on the Circumpolar list and answer the following questions:

i. Examine Zeta Ursa Majoris (the next-to-the-end star on the handle of the dipper) carefully with the naked eye, and if possible, with binoculars. What do you see?

ii. Compare the brightness of the stars in Ursa Major. What observation can you make about them?

iii. Compare the brightness of α Ursa Minoris with the second star in the handle of the "dipper." What observations can you make?

iv. If Ursa Major is lying parallel to the horizon and below the pole at 10:00 p.m. on September 14th, what will be its orientation at 4:00 a.m. on September 15th, and why?

v. Locate β Cassiopeia. This star has the proper name Caph. A line drawn through the pole star, passing through Caph, and continued southward intersects the Vernal Equinox. What is the significance of this part of the sky with respect to our seasons? (Refer to a star chart if necessary.)

(b) Constellations of Fall: Turn to face the southern horizon and the region of the sky overhead. At about 9:30 (standard time) in mid-September there are three bright stars which form the apexes of a large right triangle. The brightest of the three is Vega (in the constellation Lyra) which is farthest west, and is at the right angle of the triangle. The other two stars are Deneb (in Cygnus, the Northern Cross) and Altair (in Aquila, the Eagle). These three constellations make good reference points for the rest of the fall sky. Note in the eastern sky a large area defined roughly by four stars arranged in a great square. This is the body of the constellation Pegasus, which is connected to Andromeda.

i. Locate the stars Vega and Altair. Compare and describe their colors. What do you think their colors indicate?

ii. Locate Vega and near it the star Epsilon Lyrae. Examine this star with binoculars if possible. What do you observe?

iii. If a small telescope is available locate Beta Cygni (called Albireo). Examine the star with the telescope and report your observations.

iv. Are any planets visible? If so, identify them and locate their positions with respect to the brighter stars. Refer to a star chart to estimate the right ascension and declination of each planet. Compare the planets' brightnesses to that of the brighter stars.

v. Is the moon visible? If so, what is its phase? Locate its position with respect to the brighter stars, and estimate its right ascension and declination.

(c) Constellations of Winter: Face south or southeast at about 9:30 p.m. in January and look for a group of three stars of almost equal brightness in a line. There is a second dimmer line of three stars below and at an angle to the first line. This marks the belt and sword of Orion. This constellation makes the best reference object for the winter sky.

The belt points up and to the right toward a bright star Aldebaran in Taurus. It points down and to the left toward the bright star Sirius in Canis Major. Taurus resembles a large "V" in shape, one side of which points up to the constellation Auriga and the other side of which points to Gemini. Locate all the winter constellations listed in Table 1-I and answer the following questions:

i. Make a small table and enter descriptions of the colors of Sirius, Rigel, Betelgeuse, Capella and Aldebaran.

ii. Locate Sirius, Betelgeuse and Procyon. Describe the configuration they make and carefully compare their colors.

iii. Carefully compare the brightness of Sirius and all the other bright stars of the winter sky. What do you conclude?

iv. Use binoculars to examine the middle star in the sword of Orion. What do you see?

v. Use binoculars to examine the Pleiades. Describe your observations.

vi. Compare the brightnesses of Castor and Pollux. Which is brighter? Look up their Bayer designations and comment.

vii. Is the moon visible? If so, what is its phase? Locate its position with respect to the brighter stars and estimate its right ascension and declination.

(d) Constellations of Spring: In late spring look west at about 9:30 p.m. and locate two bright stars approximately parallel to the western horizon. These are Castor and Pollux, the two brightest stars in Gemini. Now turn south and look for a backward question mark. This marks the forequarters of Leo the lion, with Regulus as its brightest star. If you have looked at the winter sky, notice how comparatively empty the sky seems in spring.

Locate the spring stars and constellations listed in Table 1-I and answer the following questions:

i. Locate Regulus and Denebola in Leo. Compare their colors and relative brightnesses.

ii. Locate the constellation Cancer. Examine that region of the sky with binoculars. What do you see?

iii. Locate the triangle that makes up the hind quarters of Leo. Draw a perpendicular line from the hypotenuse of the triangle outward from Denebola and examine the region with binoculars. What do you see?

iv. If a small telescope is available examine α Canes Venaticorum and γ Leonis. Compare the two observations.

v. Is the moon visible? If so, what is its phase? Locate its position with respect to the brighter stars and estimate its right ascension and declination.

(e) Constellations of Summer: Locate Ursa Major and trace a line outward from the handle of the "dipper" southward until it intersects a bright yellow star. This is Arcturus in Boötis. Continue the curved line southward until it intersects a second bright (blue) star named Spica in Virgo. Now look low along the southern horizon for a fish-hook shaped figure, Scorpius. Immediately to the east of Scorpius is Sagittarius, marked by a small inverted dipper shaped group of stars. To the east of Sagittarius is a "delta" shaped grouping of stars outlining Capricornus. The four constellations Virgo, Scorpius, Sagittarius and Capricornus mark one-third of the zodiac. A fifth, Libra, can be located with some difficulty between Virgo and Scorpius.

i. Compare the colors of Spica and Antares.

ii. If you have binoculars, slowly scan the area above the "hook" of Scorpius. What do you see?

iii. Locate Corona Borealis. Compare the relative brightnesses of the stars in this constellation.

iv. If the sky is dark carefully scan the Milky Way with binoculars and record your observations.

v. Is the moon visible? If so, what is its phase? Locate its position with respect to the brighter stars and estimate its right ascension and declination.

2. STAR COUNTS

We want to determine the number of stars visible to the unaided eye using a cardboard tube. Consider a sphere whose radius is equal to the length of a cardboard tube (Figure 1-2). The area seen through the tube is equal to the circular cross-sectional area of the tube, given by

$$T = \pi r^2 = \frac{1}{4} \pi d^2 \qquad \text{Equation 1}$$

where T = area seen through tube
π = 3.1416
r = radius of the tube
d = diameter of the tube

The surface area of the whole sphere is given by

$$A = 4 \pi L^2 \qquad \text{Equation 2}$$

where A = area of the sphere
L = radius of the sphere
= length of the tube.

The ratio of these two areas will equal the ratio of the area of the sky seen through the tube to the area of the whole celestial sphere. If we use the tube to count the stars in N different areas of the sky, the total area of the sky used in counting is

$$C = N T = N \left(\frac{1}{4} \pi d^2\right) \qquad \text{Equation 3}$$

where C = total area used in counting stars.
Now we can set up the ratio:

$$\frac{\text{total number of stars counted}}{\text{total number of stars in sky}}$$

$$= \frac{\text{area of sky used in counting}}{\text{total area of sky}}$$

$$= \frac{C}{A} \qquad \text{Equation 4}$$

or

$$\text{number of stars in the sky} = \frac{A}{C} \times \binom{\text{number of stars}}{\text{counted}}$$

(a) Measure the length and diameter of your cardboard tube.

(b) Select a number of randomly distributed areas in the night sky (10 to 15 areas should be sufficient). A table of random five digit numbers may be used. Choose 10 to 15 numbers from the table; let the first three digits give the azimuth and the last two digits the altitude of each area of sky.

(c) Count the number of stars seen through the tube and record the approximate altitude and azimuth for each area.

(d) Determine the total number of stars counted, and calculate an estimation of the total number of stars visible to the unaided eye on the celestial sphere. Show all calculations.

(e) If possible, do star counts and estimate the total number of stars visible both in the city and at a relatively dark country observing site. Discuss the effects of "light pollution" on astronomical observations in general.

DISCUSSION QUESTIONS

1. What assumptions were made about the distribution of stars on the celestial sphere in using Equation 4?

2. What was the general condition of the sky while you were making observations? Was the moon visible? How might these factors affect your results? Are there any other local factors that should be considered?

3. Estimate the number of stars that might be visible to the unaided eye at a mountain top observatory on a clear, moonless night.

4. How and why might the altitudes of the areas of sky used for counting stars affect your results?

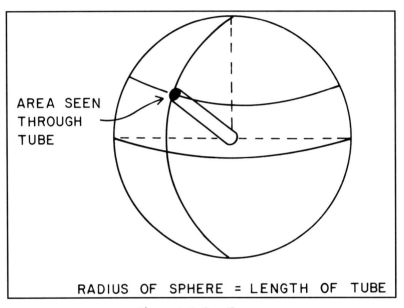

AREA SEEN THROUGH TUBE

RADIUS OF SPHERE = LENGTH OF TUBE

Figure 1-2 Star Counts.

2 Observing Exercise I: The Moon

PURPOSE AND PROCESSES

The purpose of this exercise is to observe the moon telescopically and identify some of its major features. The processes stressed in this exercise include:

Observing
Inferring

REFERENCES

1. *A Field Guide to the Stars and Planets*
2. *Norton's 2000.0*
3. *The Observer's Handbook*
4. *Sky Gazer's Almanac*

INTRODUCTION

Of all the celestial objects, the moon is still one of the most popular and spectacular to view with a small (or large) telescope. As it progresses through its phases, different areas can be observed, and the same areas can be seen in slightly different perspectives as the sunlight angle changes. This effect is most pronounced near the terminator (the sunrise or sunset line of the moon), where the shadows are longest giving the greatest light-dark contrast. The best overall viewing time is near the first or third quarter phase of the moon.

PROCEDURE

1. Set up and align the telescope (see Appendix 2) and start the clock drive if one is available. Since the moon is moving with respect to the stars, the telescope will not drive at exactly the right rate. However, use of a clock drive will allow the moon to remain in the field of view for a longer time.

2. Record the date, time, location, and general weather and sky conditions for your observations.

3. Estimate the altitude and azimuth of the moon, and indicate its position relative to nearby bright stars.

4. Observe the moon with the naked eye and do a quick sketch showing as much detail as possible. Name its phase.

5. Observe the moon with a low power (long focal length) eyepiece. Compare its markings to your naked eye sketch. How does the orientation of objects seen through the telescope compare to that seen with the naked eye?

6. Recording the focal length or magnification of the eyepiece being used, sketch the lunar surface or parts of it in as much detail as time permits. Look for and sketch as many different types of features as you can see.

7. Compare your sketch to a lunar map such as that found in *Norton's 2000.0*, a textbook, or Figures 2-1 and 2-2 and name the major maria and craters on your drawings.

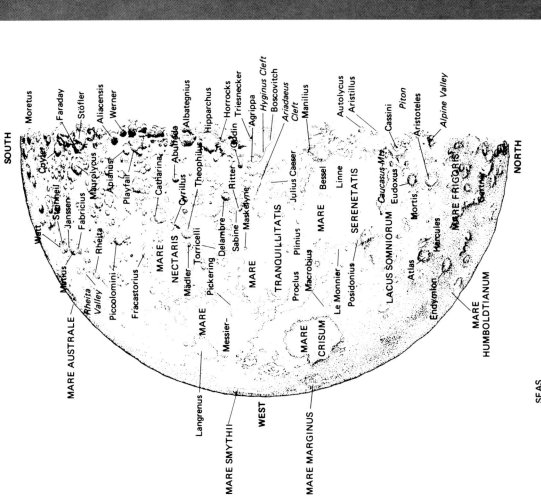

Figure 2-1. First Quarter Moon (telescopic view).

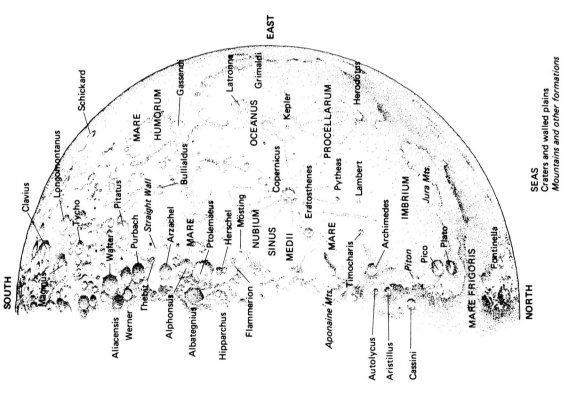

Figure 2-2. Third Quarter Moon (telescopic view).

(Photo on left: From AN INTRODUCTION TO ASTRONOMY, 2/e by Charles M. Huffer, Frederick E. Trinklein and Mark Bunge. Copyright © 1971, 1967 by Holt, Rinehart and Winston, Inc. Reprinted by permission of Holt, Rinehart and Winston. Photo on right: Mount Wilson and Las Campanas Observatories, Carnegie Institution of Washington.)

11

2. Observing Exercise I: The Moon—Observing Sheet

LUNAR OBSERVATION DATA SHEET

1) Object used for setting **RA** circle

   NAME _____ **RA** _____ h _____ m

2) Time _____ h _____ m   (AM/PM)

3) Moon's **RA** _____ h _____ m **DEC** _____ o _____

---

4) LOW MAGNIFICATION

A) Crater _____

B) Ray crater _____

C) Mountain Range _____

D) Mare _____

E) Mountain Peak _____

F) Valley _____

G) Other _____

   Eyepiece focal length _____

   Magnification _____

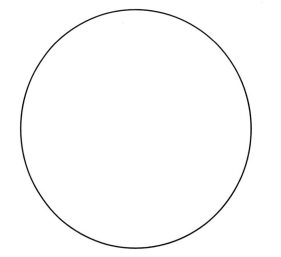

Label compass directions

5) HIGH MAGNIFICATION

Feature name _____

Description _____

_____

_____

_____

_____

_____

_____

Eyepiece focal length _____

Magnification _____

Draw a circle on the low power drawing to show the field of the high power drawing.

Constellation _____   Estimated age _____ d

DISCUSSION QUESTIONS

1. How many different types of lunar features do you see? Could you suggest some of their relative ages or possible origins?

2. Do you see more craters near the edge or near the terminator? Can you suggest several reasons why?

3. Would full moon be a good time to observe craters? Why or why not?

3 Observing Exercise II: The Planets

PURPOSE AND PROCESSES

The purpose of this exercise is to observe the planets with a telescope, and chart their positions with respect to nearby bright stars over a time period of weeks or months. The processes stressed in this exercise include:

      Observing
      Using Numbers
      Inferring
      Interpreting Data

REFERENCES

1. *The Astronomical Almanac* (current year)
2. *The Observer's Handbook*
3. *Astronomical Calendar*
4. *Sky Gazer's Almanac*
5. *Sky and Telescope* (current issue) or *Astronomy* (current issue)

INTRODUCTION

The planets generally are easy to observe and well worth the effort. Since the visible features change due to rotation or physical changes on the planetary surface itself, the planets should be observed as often as possible. Venus, Mars, Jupiter, Saturn, and occasionally Mercury are easily located with the naked eye, and Uranus and Neptune can usually be found with the aid of relatively simple star charts. Pluto, however, is only visible in a larger telescope (24-inch or larger) and requires extensive use of detailed star charts.

Positions of the planets visible with the naked eye are given monthly in *Sky and Telescope* and *Astronomy* and detailed daily coordinates for all planets are listed in the current *Astronomical Almanac*.

PROCEDURE

1. TELESCOPIC OBSERVATIONS

(a) Before coming to lab, use any of the above references to find the positions and record the coordinates of the planets.

(b) Set up and align the telescope (see Appendix 2) and start the clock drive if one is available.

(c) Record the time, date, location, and general weather and sky conditions for your observations.

(d) Estimate the altitude and azimuth, as well as the time at the beginning and end of each planet's observations. Sketch any nearby stars and locate the planet on your sketch.

(e) View the planet with the telescope starting with low magnification (a long focal length eyepiece) and then going to higher powers by selecting eyepieces with shorter focal lengths. Record the magnification or focal length of the eyepiece being used for each sketch or observation.

(f) Sketch and describe what you see in as much detail as possible. Note any visible surface markings, colorations, the general shape of the planet, locations of satellites, etc.

2. CHARTING PLANETARY MOTIONS

(a) You will need a detailed star chart (such as SC1 found in Appendix 4 and a cross-staff or plastic measuring strip. See Exercise 10, "Angles and Parallax," and Exercise 8 on "Plotting the Moon's Orbit" for information on making and using the plastic strip or cross-staff.

(b) Plot the positions of the planets with respect to the stars on the chart once or twice weekly throughout the semester or quarter. Be sure to date each recorded position.

(c) Estimate and record the brightness of each planet compared to the brighter stars for each observation.

(d) It is fairly simple to follow a planet's motion with a series of photographs using any camera on which the shutter may be left open for several minutes. See Exercise 6 "Astronomical Photography" for specific suggestions.

NAME _____   DATE _____

3. Observing Exercise II: The Planets—Observing Sheet

PLANETARY OBSERVATION DATA SHEET

1) Object used to set **RA** circle _____ **RA** _____ h _____ m

2) _____ (AM/PM)

3) Planet _____ **RA** _____ h _____ m **DEC** _____ o _____

4) **LOW MAGNIFICATION VIEW**

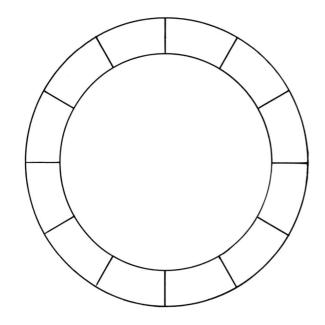

5) EYEPIECE _____ mm

    MAGNIFICATION _____ ×

NOTE: THE CIRCLE REPRESENTS THE
ENTIRE VISIBLE FIELD OF VIEW, NOT THE
SIZE OF THE OBJECT.

6) Label Compass Directions around the field of view (N, S, E, W).

7) **HIGH MAGNIFICATION VIEW**

8) EYEPIECE _____ mm

    MAGNIFICATION _____ ×

9) DESCRIPTION _____

_____

_____

_____

_____

_____

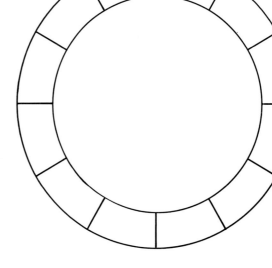

10) Label any visible moons.

DISCUSSION QUESTIONS

1. TELESCOPE OBSERVATIONS

(a) Can you think of a basis for distinguishing Jupiter's or Saturn's moons from the background stars in the field of view? What about on a longer time basis such as a few hours or days?

(b) If Jupiter is visible plot the positions of its satellites. (The four Galilean satellites are visible with small telescopes.) Consult the current issue of the *Almanac* for a section titled "Satellites of Jupiter" and identify which satellites are which.

(c) What is the phase of each planet observed? Observe or look up (in the *Almanac*) the approximate time of sunset for your location for the dates of your observations. Knowing sunset time, the time of observation, altitude of the planet, and each planet's approximate distance from the sun in A.U., use a sketch to determine the relative positions of the sun, earth, and planet in its orbit. How would you determine the planet's distance from the earth? (Hint: The sunset time and time of observation will locate the earth's HORIZON with respect to the sun. The planet's ALTITUDE will then locate it above the horizon.)

(d) What effects does the earth's atmosphere seem to have on your observations?

2. CHARTING PLANETARY MOTION

(a) In what general direction does each planet appear to move with respect to the stars? Is the direction of motion constant? If not, discuss it.

(b) Estimate each planet's angular velocity with respect to the stars in degrees/week, degrees/month, or whatever time unit seems most appropriate.

How does this velocity compare from one planet to another? Is it constant for each planet?

(c) Discuss any observed variations in planetary brightnesses.

18

4 Observing Exercise III: The Sun

The purpose of this exercise is to observe and photograph the sun's surface to study limb darkening, surface features such as sunspots and plages, and how such features change with time. The processes stressed in this exercise include:

Observing
Using Numbers
Interpreting Data
Using Logic
Inferring

INTRODUCTION

In the early 1600s Galileo first observed the sun telescopically and discovered *sunspots*. This was very disturbing to his contemporaries because the sun was thought to be a perfect, "unblemished" celestial body. Sunspots are now known to be large areas which are relatively cooler, and therefore, darker than their surrounding areas on the solar surface. They are seen to last from a few days to several weeks and are often accompanied by large outbursts of solar radiation and particles. Sunspots are observed to move across the solar disc and often disappear from one limb to reappear approximately two weeks later on the other limb.

The sun is easily observed and sunspots are usually visible in any small telescope. In addition, small bright areas, called *plages*, are occasionally seen. Watching the sun several days in a row, or two or three times a week, allows observation of sunspots forming and dying as well as their motion as the sun rotates. A good project is to observe the sun as often as possible to collect data for such studies over a semester or quarter.

BE SURE TO NEVER LOOK AT THE SUN THROUGH A TELESCOPE unless adequate precautions are taken. A solar filter may be used to block out a large portion of the sunlight (up to 99.999%) for comfortable viewing. However, some inexpensive or homemade filters don't do a sufficiently good job of blocking out the ultraviolet and infrared rays. Unless you are sure of the quality of the filter being used, it is safest to use the projection method described below.

---| PROCEDURE |---

1. SETTING UP THE EQUIPMENT

(a) Projection Method

 i. Before pointing the telescope at the sun it is necessary to stop down the aperture of the telescope by taping a cardboard ring over the end of the tube as shown in Figure 4-1. The telescope collects too much sunlight and focuses it through the eyepiece. UNLESS THIS RING IS USED TO BLOCK SOME OF THE LIGHT YOU MAY DISTORT OR MELT THE EYEPIECE! Cover the end of the finder telescope with an opaque cover to avoid the possibility of looking at the sun through it.

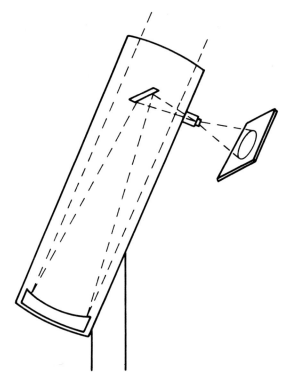

Figure 4-2. Sunspot Projection.

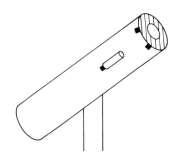

Figure 4-1. Stopping Down the Telescope Aperture.

 ii. Point the telescope toward the sun by aligning the telescope along its shadow. Hold a screen (a piece of paper or cardboard) several inches from and perpendicular to the axis of the eyepiece (Figure 4-2). Center the image on the screen and focus the sun as clearly as possible.

 iii. Hint: In order to avoid oblate solar images the screen must be placed perpendicular to the axis of the eyepiece. If photographing the image, the camera should be held with the camera back approximately parallel to the screen. It may be helpful to center the sun's image in a square drawn on the screen, and align the image in the circular pattern of the field of view of the camera.

 iv. Use a telescope drive if one is available. Although you won't be able to align it exactly on the pole, an approximate alignment will allow the sun to remain in the field for a longer period of time.

(b) Solar Filter

 i. There are two general types of filter: those that cover the end of the telescope tube, and those contained in an eyepiece. Check with your instructor for the characteristics and specific directions for the filter you will be using.

 ii. Point the telescope toward the sun by aligning it along its shadow.

 iii. If your equipment adapts for photography, attach the camera and focus it as accurately as possible.

 iv. Use a telescope drive if one is available. Although you won't be able to align it exactly on the pole, an approximate alignment will allow the sun to remain in the field for a longer period of time.

2. Note the time, date and place of your observations as well as the general sky conditions (hazy along the horizon, some scattered thin clouds, etc.).

3. Photograph the sun or sketch its disc in as much detail as possible. Show the number and positions of the sunspots. Sketch any other features that may be visible and tell what they are.

4. Observe the sunspots on as high a power as possible and sketch in detail their appearances. Indicate which spots you are observing on your diagram of the whole sun.

5. Look at an image of the sun as closely as possible and note its color or shading. Is the color uniform across the disc? Do you notice any bright areas? If so, sketch them. (A filter works best for this activity.)

6. If possible, come back a few days later and reobserve the sun, or compare your notes with someone who did the lab several days before or after you did. (Tell whose data you are comparing with.)

7. Print the best photograph for each day of solar photography. Identify the sunspots or groups on each picture or sketch, and determine the axis of solar rotation. Measure the solar latitude of each group of spots.

8. If the solar image is oblate on your prints, devise a mathematical method to correct for this in locating each group's direction and amount of motion for each time interval observed. Determine the period of solar rotation for each major sunspot or sunspot group.

4. Observing Exercise III: The Sun—Observing Sheet

SOLAR OBSERVATION DATA SHEET

1) LOW MAGNIFICATION VIEW

Date of observation _____

Time _____ (AM/PM) UT _____

Eyepiece focal length _____ mm

Magnification _____

Declination _____

Right Ascension (from SC1) _____

Zurich sunspot number R = _____

(g = _____ , f = _____).

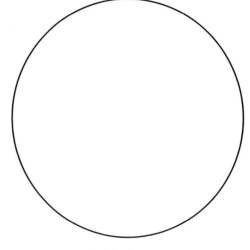

Field of View

Label compass directions in both drawings!

2) HIGH MAGNIFICATION VIEW

Eyepiece focal length _____ mm

Magnification _____

Label the umbra and penumbra of the sunspot or spot group.

Size of the spot or group in km _____ .

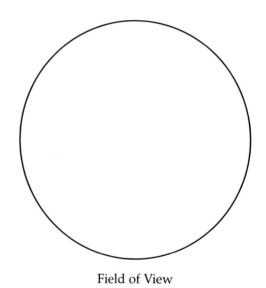

Field of View

3) Sky Conditions _____
        (i.e., clear, cloudy, hazy, windy, seeing, etc.)

4) Telescope Type and No. _____

Telescope Focal Length _____ mm

Note: Do all drawing in pencil while at the telescope. Do not attempt to work from memory and do not alter this document after leaving the telescope. Draw only what you actually see, and render it as accurately as you can.

DISCUSSION QUESTIONS

1. How do the spots move across the sun's disc over a period of a few days?

2. Is the color of the sun's disc uniform? If not, explain.

3. How did the appearance of sunspots you observed change over a time period of a few days? Is the fraction of the sun's disc traversed by a sunspot in a given time interval always the same as the spot crosses the disc? Explain.

4. Is the period of rotation the same for sunspots of different latitudes? Can you explain why or why not?

5. Do you notice any motion of sunspots relative to each other in a group? If so, explain.

5

Observing Exercise IV: Stars, Clusters, and Nebulae

PURPOSE AND PROCESSES

The purpose of this exercise is to locate and observe some double stars, star clusters, and nebulae. The processes stressed in this exercise include:

Observing
Using Numbers

REFERENCES

1. *Norton's 2000.0*
2. *Sky Atlas 2000.0*
3. *Uranometria 2000.0 Vol 1 and 2*
4. *A Field Guide to the Stars and the Planets*
5. *The Observer's Handbook*

INTRODUCTION

Double or multiple star systems, star clusters, and nebulae are among the most beautiful sights in our sky. Just look at the color photographs found in most astronomy textbooks—but remember that many of these photos were taken with long exposure times and 100- or 200-inch telescopes. Hundreds of these objects are visible in a small telescope, although it may take some patience to find some of the fainter, more nebulous ones. Others are easily seen with binoculars, and several are visible to the naked eye.

Bright double stars occasionally are given proper names but generally follow the usual naming sequence within constellations (in order of the remaining stars' increasing right ascension). In *Norton's 2000.0* double stars are often designated by their numbers in double star catalogues such as Σ 485 or BU 23. The components of a double star may also be identified "A" and "B" such as 61 Cygni A and 61 Cygni B. Most of the double stars you will be observing are bright enough to be designated by proper name or Greek letter, and can be located by using your star charts such as SC1 and SC2 (found in Appendix 4) or the finder charts in Figures 5-1 and 5-2.

Most star clusters and nebulae are designated by their Messier numbers, such as M 31, the great galaxy in Andromeda. Messier was a comet hunter and in 1781 listed over 100 nebulous objects that might be mistaken for comets. His list includes star clusters, gaseous nebulae, and what we know today to be other galaxies.

Table 5-I gives a list of suggested objects to be viewed each season. For a more complete list you might consult Messier's catalogue or a field book or atlas such as those listed in the references.

PROCEDURE

1. Set up and align the telescope and start the clock drive if one is available. Accurate alignment will be especially helpful in keeping some of the fainter objects in the field of view (see Appendix 2 "Alignment of an Equatorial Telescope").

2. Record the date, location, and general weather and sky conditions for your observations.

3. Record the time and estimate the altitude and azimuth for each object observed.

4. Locate the object with the finder telescope (accurate alignment of the finder with the main telescope is important here). Double stars should be fairly easy to spot, as most will appear double in the finder itself. Nebulous objects are often visible in the finder as small faint blurs.

5. View each object with as high power an eyepiece as practical and sketch what you see. (Note that lower powers are better for some objects as high powers tend to "wash them out" or magnify them so much that the field of view is too small.) A wide angle Erfle eyepiece is often good for larger clusters if one is available. Record the focal length or magnification of your eyepiece for each observation made.

6. Be sure to note where appropriate:

   (a) The star colors, relative brightnesses, and orientations.

   (b) The shape of the cluster or nebulous object, and the brightness distribution within it.

   (c) If you know the size of the field of view of your eyepiece, estimate the separations of the stars or the sizes of the clusters and nebulae.

Table 5-I
Objects to Observe

Name[1]	Object[2]	Constellation	Right Ascension (2000)	Declination (2000)	Comments[3,4]
All seasons					
Mizar	Double	Ursa Major	13h 24m	+54° 56′	ξ UMa, white-white, 12″
Alcor	Op Double[5]	Ursa Major	13h 25m	+55° 00′	80 UMa, with Mizar at low power
Polaris	Double	Ursa Minor	02h 31m	+89° 15′	α UMi, yellow-white stars, 19″
Summer and Fall					
Porrima	Double	Virgo	12h 42m	−01° 27′	γ Vir, yellow-white; 6″
Cor Caroli	Double	Canes Venatici	12h 56m	+38° 19′	α CVn; blue-blue, 20″
M51	Gal	Canes Venatici	13h 30m	+47° 12′	Whirlpool Galaxy
M3	Glob	Canes Venatici	13h 42m	+28° 23′	6th magnitude
M13	Glob	Hercules	16h 42m	+36° 28′	6th magnitude, bigger than M3
Graffias	Double	Scorpius	16h 05m	−19° 48′	β Sco; white-blue, 13.7″
M6	Open	Scorpius	17h 40m	−32° 13′	brightest star: yellow-orange
M7	Open	Scorpius	17h 54m	−34° 49′	good through binoculars
M20	Neb	Sagittarius	18h 07m	−23° 02′	Trifid Nebula
M8	Neb	Sagittarius	18h 04m	−24° 23′	Lagoon Nebula
ε Lyr	Double	Lyra	18h 44m	+39° 40′	two pairs separated by 208″
M57	Neb	Lyra	18h 54m	+33° 02′	Ring Nebula
M11	Open	Scutum	18h 51m	−06° 16′	resembles a globular cluster
Albireo	Double	Cygnus	19h 31m	+27° 58′	β Cyg, yellow-blue; 35″
M27	Neb	Vulpecula	20h 00m	+22° 43′	Dumbbell Nebula
γ Del	Double	Delphinus	20h 47m	+16° 08′	yellow-green (?), 10″
ξ Aqr	Double	Aquarius	22h 29m	−00° 01′	both yellowish white, 1.7″
Winter and Spring					
M31	Gal	Andromeda	00h 43m	+41° 16′	spiral galaxy, nucleus is naked-eye
Mesarthim	Double	Aries	01h 53m	+19° 17′	γ Ari, white-white, 7.8″
Almaak	Double	Andromeda	02h 04m	+42° 20′	γ And; yellow-blue, 10″
h & χ Persei	Open	Perseus	02h 20m	+57° 08′	double cluster
M45	Open	Taurus	03h 47m	+24° 07′	Pleiades, naked-eye, use binoculars
M36	Open	Auriga	05h 36m	+34° 08′	"knot of bluish diamonds"
M37	Open	Auriga	05h 52m	+32° 33′	"gold-dust," "diamonds"
M38	Open	Auriga	05h 29m	+35° 50′	"an oblique cross"
Rigel	Double	Orion	05h 14m	−08° 12′	β Ori; blue-blue, 9″
θ Orionis	Cluster	Orion	05h 35m	−05° 23′	Trapezium, four with low power
M42	Neb	Orion	05h 35m	−0° 27′	Great Nebula or the Orion Nebula
M35	Open	Gemini	06h 90m	+24° 20′	"glittering lamps on a chain"
M41	Open	Canis Major	06h 47m	−20° 44′	bright red star near center
Castor	Double	Gemini	07h 35m	+31° 53′	α Gem; white-white; 5″
M44	Open	Cancer	08h 40m	+19° 59′	Beehive cluster
Algieba	Double	Leo	10h 20m	+19° 51′	γ Leo, yellow-yellow, 4″

1. The names of some stars may vary from book to book.
2. Double = Double Star; Gal = Galaxy; Glob = Globular Cluster; Open = Open Cluster; Neb = Nebula
3. Separations of double stars are given in seconds of arc (″).
4. Quoted statements are from "Burnham's Celestial Handbook" by Robert Burnham, Jr, Dover Publications, 1978, or the "Cambridge Deep Sky Album" by Jack Newton and Philip Teece, Cambridge University Press, 1983.
5. Alcor and Mizar are naked-eye double, which may or may not be a physical binary system; separation is 12 arc minutes.

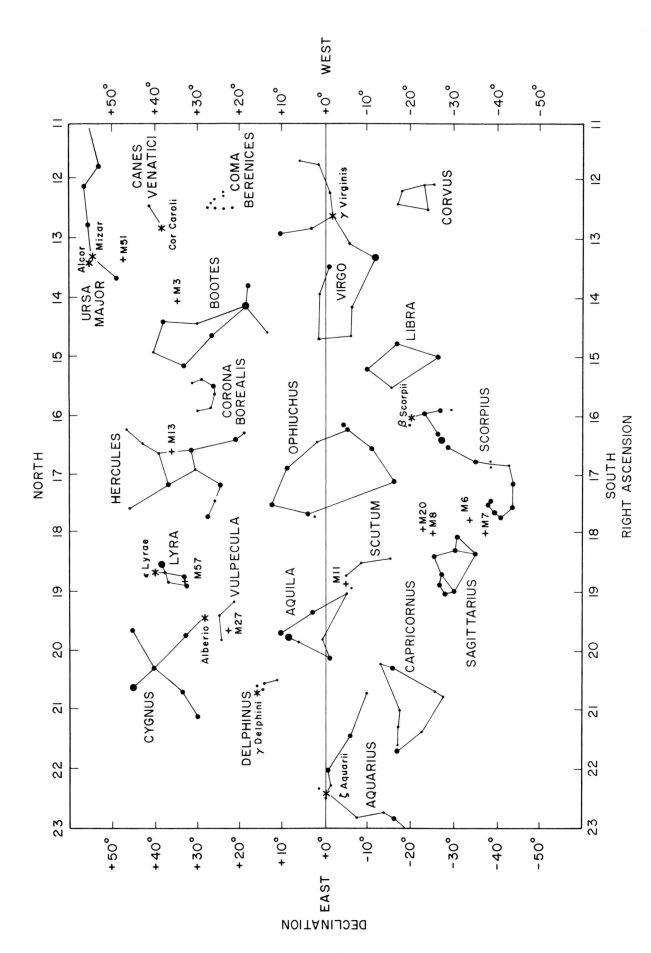

Figure 5-1. Finder Chart for Summer and Fall Objects.

28

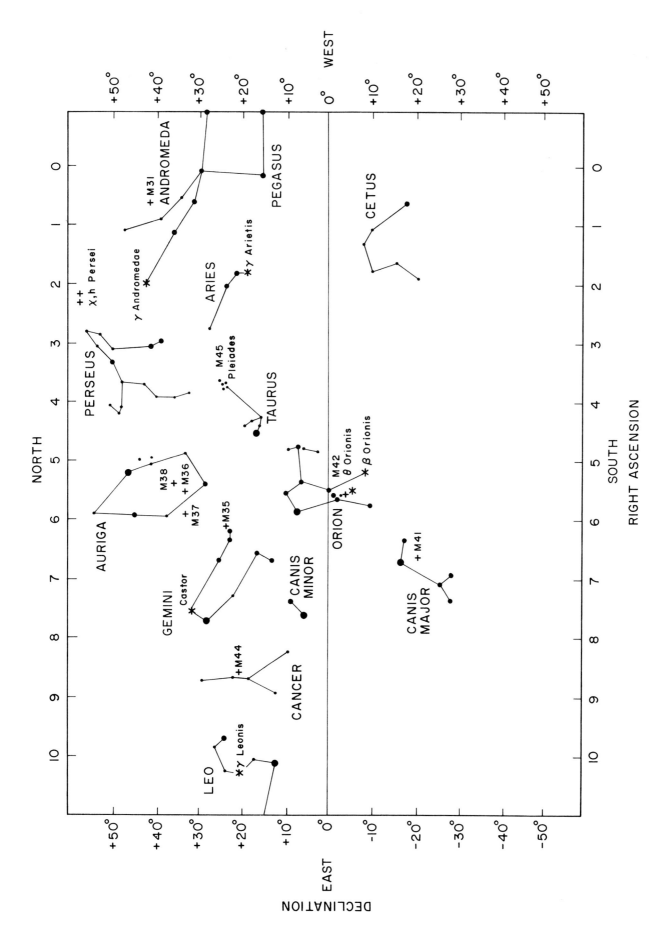

Figure 5-2. Finder Chart for Winter and Spring Objects.

29

5. Observing Exercise IV: Stars, Clusters, and Nebulae Observing Sheet

1) Circle setting star: _____ **RA:** _____ **DEC:** _____

2) Object Number: _____ **RA:** _____ **DEC:** _____

3) Name: _____ Type: _____ Constellation: _____

4) Time: _____ (AM/PM) UT: _____ **ST:** _____

5) Eyepiece: _____

   Magnification: _____

   Description: _____

_____

_____

_____

_____

_____

_____

Remember to label compass directions (N, E, S, W).

6) Object Number: _____ **RA:** _____ **DEC:** _____

7) Name: _____ Type: _____ Constellation: _____

8) Time: _____ (AM/PM) **UT:** _____ **ST:** _____

9) Eyepiece: _____

    Magnification: _____

    Description: _____

_____

_____

_____

_____

_____

_____

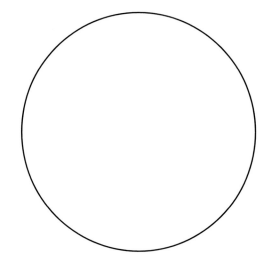

Remember to label compass directions (N, E, S, W).

6 Astronomical Photography

PURPOSE AND PROCESSES

The purpose of this exercise is to become involved in taking astronomical photographs with a Polaroid or single-lens-reflex camera or various camera-telescope combinations. The processes stressed in this exercise include:

Using Numbers
Observing
Designing Experiments
Inferring
Controlling Variables

REFERENCES

1. Eastman Kodak Company Publications
   (a) *Astrophotography Basics* (P-150)
   (b) *Scientific Imaging with Kodak Films and Plates* (P-315)
2. R. Newton Mayall and Margaret W. Mayall, *Sky Shooting, Photography for Amateur Astronomers* [New York: Dover Publications, Inc., 1968]
3. Michael Covington, *Astrophotography for the Amateur,* [London, England: Cambridge University Press, 1985]

INTRODUCTION

Photography has contributed extensively in advancing the field of astronomy. The introduction of photographic techniques in the late 1800s expanded some areas of research and made many new areas possible. For example, about seventy years of work in determining stellar parallaxes had yielded values for only about 55 stars. In 1904 Schlessinger applied photography to the field and in the subsequent seventy years trigonometric parallaxes have been determined for about 6000 stars.

The earliest known astronomical photograph was made in 1840 by Henry Draper (a New York physician) when he completed a successful 20-minute exposure of the moon. In 1845 Fizeau and Foucault obtained the first known photo of the sun, and in 1850 William Bond secured the first stellar photographs of Vega and Castor. In time, improved guiding and photographic techniques have made photography an integral part of astronomy.

Astronomical photography can be done for pleasure as well as for scientific purposes. A good picture does not necessarily require expensive equipment, but can often depend on the skill, ingenuity, and luck of the photographer. This exercise describes a variety of photographic techniques that the beginning student can use. Nearly any camera with a variable shutter speed can be used, but for serious work it is best to have a single-lens-reflex camera (SLR) with a removable lens. Most cameras of this type have a "time" or "bulb" exposure setting and can be attached to a tripod.

$\boxed{\text{PROCEDURE}}$

The techniques involved in this exercise include (1) choice of film types, (2) use of a Polaroid camera, (3) use of SLR cameras, and (4) camera and telescope combinations. Try as many types of photography as time and equipment permit. Do not be afraid to try some of your own ideas: do what interests you! Preserve records of your exposures! (See Appendix 4 for a recording sheet.)

1. CHOICE OF PHOTOGRAPHIC FILMS

An important first step in obtaining good photographs is choosing a film type suitable for your subject matter. Several common Eastman Kodak films work well for astronomical photography. Their general emulsion characteristics are described below:

(a) Black-and-White Films

    i. Tri-X (ASA 400)[1]

    This film is good for indoor and some outdoor work. However, it may be too "fast" for direct sunlight. It is good for astronomical work because of its speed, but it is somewhat grainy.

    ii. Plus-X (ASA 125)

    This film must be used outdoors or with a flash. Though its speed is low (requiring a longer exposure time), it has a fine grain. Thus it is suited for astronomical photography of bright objects, as prints may be greatly enlarged without appreciable graininess.

    iii. 103aF (no ASA)

    This film, used primarily for astronomical work, has a very low reciprocity failure[2] (thus no ASA rating) and can therefore be used for long time exposures. It also has the advantage of sensitivity over a large range of wavelengths. However, this film is *very* grainy.

(b) Color Films

    i. Ektachrome 200 (ASA 200)

    ii. Ektachrome 64 (ASA 64)

    These two types of film are similar in that they are both good general-purpose films. However, the Ektachrome 64 is better suited for astronomical work (especially for planets) as it has a fine grain.

2. POLAROID PHOTOGRAPHY

A Polaroid camera provides an excellent method of becoming acquainted with astronomical photography. The ease in obtaining finished pictures often offsets the comparatively low quality of images obtained. In order to use a Polaroid for astronomical work you must have a way to attach the camera to a tripod. Many Polaroids are not equipped with an accessory base; however, one can be constructed or purchased locally and glued with epoxy to the bottom of the camera. The Polaroid must be equipped with either a "time" setting or an electric eye since it is necessary for the shutter to remain open for extended periods of time. If the camera does not have a "time" setting but does have an electric eye it is possible to "fool" the shutter into staying open by covering the electric eye with a piece of opaque tape. The shutter will then stay open as long as the shutter button is depressed.

To use the camera attach it to the tripod and cover the electric eye if necessary. Adjust the distance setting to infinity (or its greatest distance) and have the lens aperture open as far as possible. Use black-and-white film, but set the camera film selection indicator at the "color" setting to insure maximum aperture. Some suggested projects using the Polaroid include:

(a) Constellations: Photograph a number of different constellations using about two-minute exposure times. Determine the limiting magnitude and plate scale in degrees per cm for each exposure by consulting an atlas or star chart.

(b) The Moon: Photograph the moon's position on two successive evenings and determine its approximate period by determining its angular motion in the sky. (You might also wish to consult Exercise 16 "The Moon's Sidereal Period.") You will need to use the angular separation between two known stars in order to determine a plate scale. A one-minute exposure will slightly overexpose the moon but will pick up the brighter stars in the field.

(c) Circumpolar Star Trails: Complete a 20-minute circumpolar exposure and determine the earth's rotational period. (You might wish to consult Exercise 13 "Length of the Sidereal Day.") We suggest that exposing the picture for 15 minutes, interrupt the exposure for three minutes by covering the lens, and then

---

1. ASA is an exposure index given by the American Standards Association. Many light meters are calibrated in terms of ASA exposure indices. Currently film manufacturers are designating film speeds with a two part I.S.O. number. (International Standards Organization.) The first number (e.g., ISO 400/28°) is the ASA number.

2. Reciprocity failure refers to the fact that the response of most films is usually not linear with time for exposures of over a few seconds duration. The fact that 103aF is a low reciprocity failure film can mean that in order to photograph stars one magnitude fainter, the exposure time must be increased by a factor of three. However, for high reciprocity failure films such as Tri-X or color film, increasing the exposure time by a factor of three or four may allow you to record stars only a half magnitude fainter.

expose the film for an additional two minutes. From this you can determine the direction of rotation as well.

(d) Planets: Photograph a planet near a prominent constellation several times over a two-week or longer period. Determine its direction of motion and period relative to the fixed stars using the angular separation between identified stars in order to determine a plate scale. A two-minute exposure works well.

3. SLR CAMERAS

Any camera with a "time" or "bulb" setting can be used for photography in the same manner as the Polaroids described above. However, if you have access to a SLR camera with a good quality lens you have the advantage of being able to take shorter exposures without loss of image quality. High speed black-and-white film such as Tri-X or color film such as Ektachrome 200 or 400 are recommended. (Ektachrome films can be processed at ASA 400 or 800 respectively if requested.) For all of the activities suggested in this section it is assumed that an ordinary 50-55 mm lens is being used, that the camera focus is set at `infinity,' that the aperture is set at its maximum opening, and that a sturdy tripod is available. In addition, use of a cable release will decrease the amount of camera vibration. A suggested technique for long time exposures is to hold a piece of opaque material over the lens (but not touching it) until the cable release is depressed and locked; then remove the material to begin the exposure. When the picture is completed reverse the process.

(a) Constellations: Excellent constellation slides can be obtained using Ektachrome films and 30-second to 1-minute exposures. A dark sky away from city lights enhances the quality of the slides. The 50–55 mm lens produces image sizes that will encompass all but the largest constellations (such as Draco or Eridanus).

(b) Star Trails: Simply point the camera at either the polar region or toward the celestial equator and allow the shutter to remain open for about 30 minutes. Contrast the shape of the star trails and note that Polaris is not precisely at the rotational pole. Contrasting stellar colors show up well on color slides and differences in magnitudes become very obvious. If the sky is extremely dark longer exposures can be made.

(c) Comets: If a bright naked eye comet is available excellent photographs can be made with either black-and-white or color film using exposure times as short as 30 seconds. Photograph the comet on two successive nights to show its comparatively rapid motion among the stars. If possible it is nice to include part of the horizon to border the photograph and provide a general scale.

(d) Aurora: Recommended exposure times for aurorae range from 10 seconds to two minutes with Ektachrome films. Excellent auroral pictures with Tri-X film have been made with exposure times as short as five seconds. If possible, try to photograph an aurora against a prominent constellation for added effectiveness.

(e) Meteors: Meteor photography can be frustrating as meteors are elusive objects. Although five sporadic meteors per hour is a commonly observed rate, you can never be certain in which direction your camera should be pointed. The best chance of success in photographing a meteor trail is during one of the predicted meteor showers. Point the camera about 30–45 degrees away from the radiant point and leave the shutter open for 15 to 45 minutes. If a meteor is seen in the direction the camera is pointing, close the shutter and advance the film to the next exposure. Refer to the photographs in Figure 18-1.

(f) Bright Artificial Satellites: Information on passage times of bright satellites can be obtained by using a computer program designed to plot the ground paths of satellites. If possible, pick a time when the satellite will pass through a prominent constellation. Open the aperture to its fullest extent and hold the shutter open for a measured length of time. If the satellite trail and the constellation appear clearly on the negative, make a print and devise a plate scale by using the angular separation of two bright stars. Determine the period of the satelite's orbit.

(g) Planetary Conjunctions and Alignments: Occasionally two or more planets are seen in the same area of the sky providing an excellent subject for photographs. If the alignment takes place near the sun (as always is the case for Mercury and Venus) the planets can be photographed just before dawn or right after sunset. By including the horizon in the picture you can determine the inclination of the ecliptic with respect to the horizon. If three planets are visible, see if all three planets fall in the same orbital plane. Five to twenty-second exposures work satisfactorily with color film.

4. CAMERA-TELESCOPE COMBINATIONS

A camera can be used in several ways with a telescope. Success with any of the methods is dependent on the equipment available and the care and patience exhibited by the photographer.

(a) The telescope can be used as a guiding device for an ordinary camera. The camera is attached to the side of the telescope tube and the telescope's equatorial drive keeps the camera pointing at the same region of the sky. For extended exposures an illuminated cross hair can be used with the telescope eyepiece to make

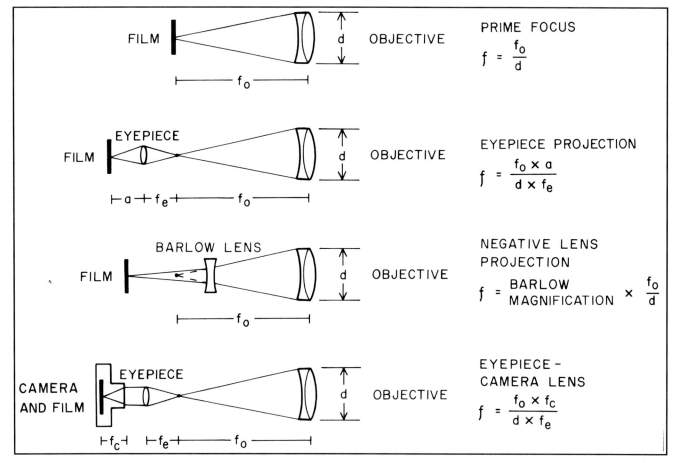

Figure 6-1. Camera-Telescope Combinations.

minor adjustments to telescope guiding. Using a tele-photo lens, it is possible to photograph bright aster-oids and planets Uranus and Neptune with Tri-X film and exposure times of one to ten minutes. Prominent clusters such as the Beehive, h and χ Persei and Pleiades nicely fill the 35 mm frame with 135 to 200 mm lenses (use 3–5 minute exposure times). With practice you should be able to locate M 31 and the Ring Nebula on this type of exposure.

(b) Four methods can be used to attach a camera to a telescope to take advantage of the telescope's ability to magnify as well as to track celestial objects. An ef-fective f/ratio can be calculated for each, using Figure 6-1. Suggested exposure times for a variety of celestial objects can be estimated from Table 6-1. It is a good idea to bracket these times with exposure times on either side of the suggested value in order to account for differences in seeing conditions and the cleanliness of the optics of your system.

The four different camera-telescope combinations are the Eyepiece-Camera Lens system, Eyepiece Projec-tion, Prime Focus, and Negative Lens Projection.

i. The Eyepiece-Camera Lens System: Insert the eyepiece into the telescope, leave the lens on the camera, and mount the camera on a bracket over the eyepiece. (Brackets for this purpose are com-mercially available or can be constructed locally.) Place a piece of tissue paper over the end of the telescope tube and illuminate it with a 100 watt bulb from about two feet away. Set the camera lens at its smallest aperture. Move the camera in and out on the bracket until the field of view ap-pears evenly illuminated. Remove the paper, focus on a bright astronomical object, open the lens to its largest opening, and set the distance to infinity. Use the telescope eyepiece to focus the object.

ii. Eyepiece Projection: This is similar to the method given above except no lens is used on the camera. Remove the lens from the camera and mount the camera on the telescope bracket directly over the eyepiece opening. Look through the camera and focus the object by mov-ing the eyepiece outward slightly from the set-ting used for visual observation.

Table 6-1

N	1	2	3	4	5	6	7	8	9	10	11	12
f/#	2.0	2.8	4.0	5.6	8	11	16	22	32	45	64	90
ASA	16	32	64	125	250	500	1000					

Exposure Step Change = n (f/ratio) – N (ASA)

Exposure Time (s)	Object to be Photographed
1/4000*	Full Sun+
1/2000*	Venus (greatest elongation); Solar Prominence; 3/4 Solar Eclipse+
1/1000	Diamond Ring Effect
1/500	Jupiter
1/250	Full Moon
1/125	Solar Corona; 3/4 Lunar Eclipse
1/60	Mars; Quarter Moon
1/30	Saturn
1/15	Crescent Moon
1/8	
1/4	
1/2	
1	
2	Galilean Satellites
4	
8	
16	Total Lunar Eclipse

*Not on most cameras.

+Exposures are calculated for a Number 4 neutral density filter.

iii. Prime Focus: This system produces the sharpest images and is photographically the "fastest" because it uses the fewest optical elements. In order to photograph objects with small reflectors at prime focus an adapter that replaces the camera lens is required. (Most camera stores can provide them.) Such adapters usually come in two parts: an adapter ring which directly replaces the camera lens and a T-adapter which has an outside diameter equal to that of the telescope eyepiece. The primary telescope mirror will have to be moved forward a short distance in the tube because the image plane of a reflecting telescope is not directly accessible. Some vignetting will occur because of the tube diameter and the size of the secondary mirror. For most astronomical purposes, however, this effect will produce no serious problem.

The camera and attached adapter is simply slid into the eyepiece holder and the telescope rebalanced to accommodate the weight of the camera. The object is focused using the view finder of the camera. As in the other combinations used above, the comparatively low level of illumination of celestial objects requires great care in focusing.

iv. Negative Lens Projection: A more common system employed with small reflectors is the negative lens system. Negative lenses (also called Barlow lenses) make it possible to extend the focal length of the system and effectively "pull" the image out of the eyepiece tube without repositioning the primary mirror. As a result this method provides greater magnification. The camera and telescope are set up as in the prime focus system described above with the negative lens used in addition to the adapters.

Barlow lenses are usually available from optical supply houses. Solar projection filters also serve as Barlow lenses and this system can be used for sunspot photography. Exposure times or the camera's built-in light meter can be used if one is available.

5. PREDICTION OF EXPOSURE TIMES

The choice of exposure times for various astronomical objects is largely an empirical process involving the f/ratio of the system, the film type used, and the object chosen to be photographed. Table 6-I[1] presents data that permit a quick estimate of suggested exposure times. To use the Table:

(a) First calculate the f/ratio of your camera-telescope system by using Figure 6-1.

(b) Locate this f/ratio and the ASA of the film to be used on the upper part of Table 6-I. If these two numbers are directly above and below each other, locate the object to be photographed in the lower part of the Table and use the listed exposure time. It is good to "bracket" this suggested time by one or two exposures faster or slower to account for possible unusual sky conditions or optics which might not be perfectly clean.

(c) If the two values (f/ratio and ASA) do not appear directly above and below each other we must calculate the exposure step change. Find the N values directly above the f/ratio and ASA at the top of the Table. The correction is given by the algebraic difference of these N values where:

N (f/ratio) – N (ASA) = exposure step change.

(d) If the correction is positive find the exposure time by counting downward from the exposure time listed with the object by a number of steps equal to the difference; if the number is negative, count upward the appropriate number of steps.

For example, suppose that an f/8 telescope-camera system and Plus-X film are to be used to photograph Jupiter.

   i. f/8 gives N (f/ratio) = 5

  ii. Plus-X film, ASA 125 gives N (ASA) = 4

 iii. Exposure step change = 5 – 4 = +1.

 iv. The suggested exposure time is found by counting downward one step from the Table value for Jupiter of 1/500 s to 1/250 s.

---

1. Exposure table and procedure developed by Mark Hodges, Owens Valley Radio Observatory.

6. Astronomical Photography

Photography Log Sheet

NAME _____   FILM TYPE _____

DATE _____   ROLL NUMBER _____

Lens System	Frame	Exposure Time	Object
	1		
	2		
	3		
	4		
	5		
	6		
	7		
	8		
	9		
	10		
	11		
	12		
	13		
	14		
	15		
	16		
	17		
	18		
	19		
	20		

7 Field of View of a Telescope

The purpose of this exercise is to determine the angular size of the field of view of a telescope and to study the dependence of the rate of drift of a star through the field of view on the star's declination. The processes stressed in this exercise include:

Observing
Designing Experiments
Using Numbers
Controlling Variables
Inferring

INTRODUCTION

Since the earth has a rotational period of about twenty-four hours, a star will travel one twenty-fourth of the way around its small circle in one hour of time. If we place a star in the field of view of our telescope, we can measure how long it takes to drift across the field, and convert this measure of time into an angular measurement. However, the time it takes a star to traverse a given field of view depends on the star's declination: a star near one of the celestial poles will appear to move more slowly through the field of view than a star at the equator. This occurs because the earth has the same period of rotation with respect to any star. However, a star nearer the pole has less distance to go along an imaginary surface of the celestial sphere (in Figure 7-1, r_2 is less than r_1) and thus its rate of drift will be slower.

The celestial equator is the only circle of constant declination that will give transit times directly equal to the arc measure of the field of view of the telescope. However, a mathematical relationship exists between the declination of a star and its rate of drift.

PROCEDURE

1. Your instructor will assign each student or group of students a star for which you are to determine the rate of drift. Center this star in the field of view with the drive still going and adjust the telescope so that the star will traverse the widest part of the field.

2. Set the telescope field just to the west of the star using the slow motion control in right ascension if one is available. This puts the star barely out of the field of view such that it will drift through the center of the field when the drive is shut off. Be careful during this maneuver not to change the declination setting, or the star will no longer traverse the center of the field.

3. Turn off the telescope drive. Have a stopwatch ready so you can accurately measure the star's transit time. Start the watch when the star first appears, and stop it when the star disappears from view.

4. Repeat this procedure at least three times for each star and average the results.

5. Each group of students will measure the rate of drift for a star of different declination. At the end of the laboratory period all the data will be compiled. Find the mathematical relationship between a star's declination and its rate of drift.

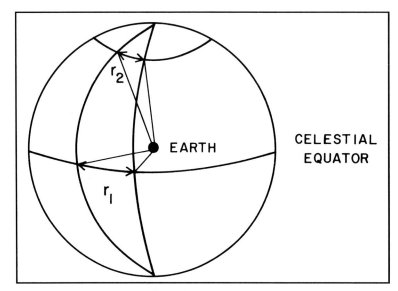

Figure 7-1. Small Circles at Different Declinations.

6. Determine the angular field of view of the telescope, using the conversions:

   1 hour of time = 15 arc degrees
   1 minute of time = 15 arc minutes
   1 second of time = 15 arc seconds

7. Repeat this procedure for other eyepieces used with your telescope.

DISCUSSION QUESTIONS

1. Using your value for the field of view of the telescope, how long would it take for the moon to go from first to third contact? Explain.

2. An observer wishes to use your telescope to observe α and β Lyra simultaneously. Will this be possible? Explain.

8 Plotting the Moon's Orbit

PURPOSE AND PROCESSES

The purpose of this exercise is to acquaint the student with the fundamental motions of the moon, the concepts of the ecliptic and celestial coordinate systems, and to familiarize him with some of the constellations of the zodiac. The processes stressed in this exercise include:

Observing
Using Numbers
Inferring
Prediction
Formulation of Models

INTRODUCTION

The moon is one of the most prominent objects in the sky, and nearly everyone is familiar with its phases. Yet few people study it critically to observe exactly how it moves among the stars, or to learn how position relates to lunar phase. Such a study also provides an excellent opportunity to become familiar with the constellations and some of the brighter stars.

PROCEDURE

1. Examine the SC1 Constellation Chart found in Appendix 4. Become familiar with the constellations on this chart and its system of depicting stellar magnitudes. Be certain that you understand the significance of the ecliptic, the celestial equator, right ascension, and declination.

2. Observe the moon with respect to the brighter stars as often as possible for the next four to eight weeks. Plot the position of the moon on the star chart, and record the date and time of each observation.

In order to estimate angles between the moon and brighter stars a cross-staff or calibrated plastic strip should be used. Calibration and use of the plastic strip is described in Exercise 10 "Angles and Parallax." A cross-staff can be constructed from a meter or yardstick and a sliding, perpendicular cross-piece (Figure 8-1).

Hold the cross-staff near one eye and slide the cross-piece along the staff until one end is on a bright star and the other is on the moon. The angle between the two can be calculated using the simple trigonometric relations given in the figure.

Measure the angle between the moon and at least two (or preferably three) nearby stars. Be sure to record the angles and names of the stars being used (a log sheet for your observations is provided in Appendix 4). These readings may be used to triangulate the moon's position on your star chart. As an example, say we have the data given in the first two columns of Table 8-I:

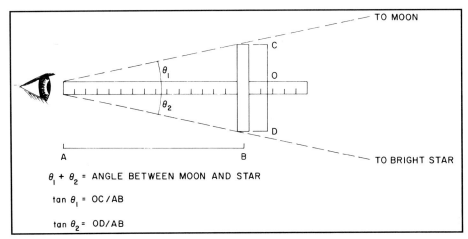

Figure 8-1. A Simple Cross-Staff.

Table 8-I
Sample Moon Plotting Data

Star	Angle	Radius on SC1
Procyon	12°	2.0 cm
Pollus	13°	2.2 cm
Betelgeuse	20°	3.4 cm

(a) Look at the declination scale on the star chart and determine the scale factor in cm per degree. Convert your degree measurements to cm.

(b) If the moon is 20° from Betelgeuse then it must be located somewhere on a circle of 20° radius centered on Betelgeuse, and likewise for the other stars.

Using a compass, mark off an arc of the radius determined above in the general direction of the moon for each star (Figure 8-2).

(c) The only way the moon can simultaneously be 12° from Procyon, 13° from Pollux, and 20° from Betelgeuse is for it to be at a point where the arcs for the three stars intersect. Find the point closest to where the three intersect (due to error or uncertainty in measurement they may not intersect exactly) and plot the moon there.

3. From your completed chart and data, answer the following questions:

(a) In what direction does the moon travel in its orbit?

(b) What is the approximate motion of the moon in 24 hours (expressed in degrees)?

(c) What is the right ascension of the ascending node of the moon's orbit?

(d) What is the inclination of the moon's orbit to the ecliptic? (Show on your chart how this is determined.)

(e) What time does the third quarter moon rise? Explain.

(f) What time is the full moon overhead? Explain.

(g) What time does the first quarter moon set?

4. The age of the moon (which accurately gives its phase) starts at zero each time the moon is new. From a calendar or almanac find the day and approximate time when new moon occurred during your period of observation.

(a) What is the age of the moon (in days) at the ascending and descending nodes of the orbit?

(b) What is the significance of knowledge of lunar age at the nodal points? Explain.

| **OPTIONAL** |

Each time you observe the moon, look carefully (without optical aid) at the lunar surface features. Make a sketch of the lunar disc. There are at least five maria that can be observed. Please do not refer to lunar maps but attempt to complete these sketches as the ancients could have done. Interestingly enough, there are no recorded attempts at this task before telescopes were available in the early 17th century. It is reported that the early telescopic charts are strangely inaccurate.

Can you draw any conclusions about the rotation of the moon from your sketches?

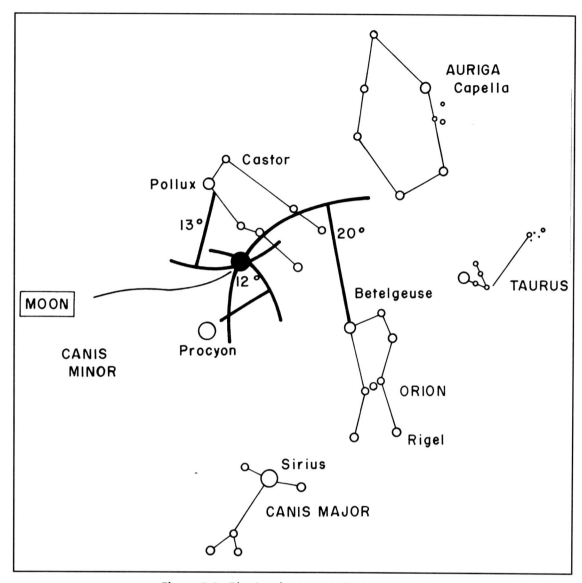

Figure 8-2. Plotting the Moon's Position.

8. Plotting the Moon's Orbit

Data Log Sheet for Plotting the Moon's Position

Date	Time	Star 1	Angle	Star 2	Angle	Star 3	Angle

9 Star Charts and Catalogues

PURPOSE AND PROCESSES

The purpose of this exercise is to become familiar with some of the more common star charts, catalogues, pathfinders, and timetables. The processes stressed in this exercise include:

Identifying Variables
Prediction
Controlling Variables
Using Logic
Inferring

REFERENCES

The following are star charts, finders, timetables, and catalogues particularly suited for naked eye observation or for use with a small telescope:

1. SC1 and SC2 or similar Constellation Charts
2. *Sky Gazer's Almanac*
3. A Planisphere or Star and Satellite Pathfinder
4. *The Astronomical Almanac* (current year)
5. *Norton's 2000.0*
6. (a) *Sky Atlas 2000.0*
   (b) *Uranometria 2000.0*
7. *Sky Catalogue 2000.0 Volumes 1 and 2*
8. *Burnham's Celestial Handbook Volumes 1, 2, and 3*

INTRODUCTION

Star charts and catalogues have been compiled since astronomical events and observations were first recorded. Today there are many devices on the market for locating stars, but here are some of the ones you're most likely to run into (how to order them is given in Appendix 3 "Equipment Notes").

Since this exercise describes the use of several different types of star charts and observing aids, be sure to consult your instructor to determine which parts to do.

1. SC1 AND SC2 OR SIMILAR CONSTELLATION CHARTS
Most constellation charts are designed to be used outdoors to represent the sky and aid in actually finding the desired objects. They give right ascensions and declinations, indicate approximate magnitudes, and depict the general pattern the stars in a constellation appear to make.

2. *SKY GAZER'S ALMANAC*
This almanac is a graphic plot of local rising, setting and transit times of planets, the sun, the moon, and other bright objects for each date of the year. It also indicates the brighter meteor showers and the phases, apogee, and perigee of the moon.

3. PLANISPHERES AND PATHFINDERS
These devices are also designed for use outdoors but usually contain less detail than star charts. They have the advantage of adjusting for your date and time to aid in orienting a star chart with respect to your local horizon.

4. *THE ASTRONOMICAL ALMANAC*
This reference, published yearly, lists daily positions for the planets, moon, bright asteroids, and the sun; gives data on planetary satellites, eclipses, sunrise, sunset, moonrise, and moonset; lists coordinates for the brighter stars; and gives general data on the solar system. It is often used at the telescope as well as for general reference purposes.

5. *NORTON'S 2000.0*
This combination catalogue and atlas is designed primarily for use by amateur astronomers. It contains detailed constellation charts (good for telescopic observing) and lists data on objects of interest for the naked eye and small telescope.

6. (a) *SKY ATLAS 2000.0*   (b) *URANOMETRIA 2000.0*
Both are collections of sky maps showing stars, clusters, nebulae and galaxies for the epoch 2000.0. The *Sky Atlas* has 26 charts with stars to 8.0 magnitude in either black and white charts or color spiral-bound book. The *Uranometria* has 259 charts with stars to 9.5 magnitude and includes the positions of quasars and radio and X-rays sources in a hard bound book.

7. *SKY CATALOGUE 2000.0 VOLUMES 1 AND 2*
This set lists the general star data for stars of magnitude 8.0 and brighter (Vol. 1) and for double stars, variable stars, and nonstellar objects (Vol. 2). Edited by Alan Hirshfeld and Roger W. Sinnott, the books are available in both paper and hard back from Sky Publishing.

8. *BURNHAM'S CELESTIAL HANDBOOK VOLUMES 1, 2, AND 3*
These books contain similar data to that of all the above and more. The brighter stars, clusters, nebulae, and galaxies in each constellation are discussed in both mythical, historical, and physical detail. Some of the data may be out of date considering the books were published in 1978. A must for people who like trivial details.

PROCEDURE

---

1. CONSTELLATION CHARTS
Two charts, SC1 and SC2, are in Appendix 4 at the back of this manual. SC1 represents the equatorial region of the sky and SC2 the north polar region.

(a) First look at SC1. The heavy line across the center of the chart denotes the *celestial equator*. The wavy line intersecting it represents the apparent path of the sun during the course of a year (called the *ecliptic*). Along the side of the map are numbers running from 0 at the equator to $\pm 60°$ at the top and bottom. These are *declination* (dec) markings and are analogous to latitude on the earth. They stop at 60° rather going on to 90° to avoid distortion of the map near the poles. The coordinate going across the top and bottom of the map is called *right ascension* (RA). Since the entire sky goes over our heads once a day, it is measured more conveniently in hours than degrees. 24 hrs = 360° so that $1^h = 15°$.

(b) Locate the following objects and record their approximate right ascensions and declinations. Tell whether each is a star or constellation.

1. Arcturus   3. Aries   5. Orion
2. Sagittarius 4. Sirius

(c) The small dates along the sun's path or ecliptic tell when we "see" the sun in that position in the sky. Find and record the coordinates of the sun for today. What constellation is the sun in?

(d) At midnight the sun is halfway (or 12 hours) around the sky from our north-south overhead line (called the *meridian*). Thus, stars located 12^h from the sun will be on the meridian at midnight. Give the RA of the meridian for midnight tonight and list several stars and/or constellations in that area. (Note: Star charts are set up for standard time. If you are presently observing daylight time it will be easiest if you do this exercise for standard time, and convert to daylight time as a last step if necessary for local observing.)

(e) Since we see the stars moving from east to west during the course of an evening, a star 1 hour WEST of the meridian at midnight should have crossed the meridian one hour before, at 11:00 P.M. Note that on this chart RIGHT IS WEST AND LEFT IS EAST as you look down on it. This is because it is made to be held over your head as you face south. Using standard time, give the RA of the meridian at 11:00 P.M., and name a star or constellation that will be there then. Repeat for 8:00 P.M. and 2:00 A.M. Check your 8:00 P.M. results with the dates along the bottom of the map.

(f) This process may also be used in reverse to find what date a given star will be in a good observing position. Follow the example below.

Example
On what date will the star Vega be on the meridian at midnight, and when at 10:00 P.M.?

1. Find Vega: RA = $18\frac{1}{2}^h$.

2. At midnight the sun is 12^h away from the meridian, so $RA_{sun} = 18\frac{1}{2}^h - 12^h = 6\frac{1}{2}^h$.

3. Look at the ecliptic at RA = $6\frac{1}{2}^h$. The date is June 30.

4. Vega will be on the meridian at midnight on June 30.

5. Since the stars all rise and set 2 hours EARLIER each month, it will be on the meridian at 10:00 P.M. one month later, or on July 30.

(g) On what date will Arcturus be on the meridian at midnight? At 2:00 A.M.? Show your work!

(h) Now look at SC2. It shows the north polar region of the sky, and Polaris is at the center. The concentric circles show declination, and right ascension lines run outward like the spokes of a wheel. The best way to use this chart outdoors is to hold it over your head, find the Big Dipper or Cassiopeia, and rotate the chart until it resembles the sky when facing north.

(i) Note that each right ascension line has a date with it. This is the date that the stars are on the meridian ABOVE the pole at about 8:00 P.M. local standard time.

On what date will Caph (in Cassiopeia) be on the meridian at midnight? Alcor and Mizar (in Ursa Major) at 10:00 P.M.? Show your work.

2. *THE ASTRONOMICAL ALMANAC*

(a) Listings in the *Almanac* are given for Universal Time, or the time in Greenwich, England, to avoid local time zone confusion. (The longitude of Greenwich = 0°.) Universal Time is also broadcast on WWV shortwave radio for navigators and astronomers.

To convert from local to Greenwich time you need to know your longitude. For example, if you are at approximately 90° W longitude you are 6 time zones west of Greenwich. Therefore, it is always 6 hours LATER in Greenwich than at your location. For example, 10:00 (local standard time) + 6 hours = 4:00 A.M. TOMORROW in Greenwich. Thus, for 10:00 P.M. here, you will want to look under 4:00 A.M. on tomorrow's date for the moon's listings. For most purposes the planets move little enough from hour to hour that only daily listings are given.

Record your longitude and determine what Universal Time corresponds to 10:00 P.M. your local standard time.

(b) Locate and record the right ascension and declination of the moon and five naked eye planets (Mercury, Venus, Mars, Jupiter, and Saturn) for 10:00 P.M. tonight. Watch the signs on declination, as they are not given for each listing in the table.

(c) Plot the positions of the moon and planets on your star chart. Which constellation is each in?

(d) Which planets will be visible at 10:00 P.M. tonight? In what general part of the sky will they be seen? Remember that for a northern mid-latitude observing location you can see approximately 6 hours on either side of your meridian at the equator, slightly more at northern declinations, and slightly less at southern declinations.

3. PLANISPHERES OR STAR AND SATELLITE PATHFINDERS

The general directions for using most planispheres and pathfinders are given on the back. It is usually faster to determine where things are and what part of the sky you're looking at with one of these devices. However, they are not as accurate or detailed as your star charts, and the constellations are often easier to recognize on other charts.

(a) Set up the planisphere or pathfinder for today at 11:00 P.M. standard time. Name several stars and constellations that will be on the meridian. Compare these with your results in part 1(e).

(b) On what date will Arcturus be on the meridian at midnight? At 2:00 A.M.? Compare these with your results in part 1(g). When will Caph be on the meridian at midnight? Alcor and Mizar at 10:00 P.M.? Compare to part 1(i).

(c) On what date will each of the following be found on the meridian at the times indicated?

    i. Andromeda at 10:00 P.M.

    ii. Andromeda at midnight.

iii. Orion at 8:00 P.M.

iv. Sirius (in Canis Major) at 10:00 P.M.

4. *SKY GAZER'S ALMANAC*

(a) The almanac is set up for a latitude of 40° N and a longitude in the center of a time zone. For many observers this will be close enough to your location that corrections need not be used. Check with your instructor; if corrections are necessary, they are given in the explanatory notes that come with the almanac.

(b) This timetable contains more information than most people ever need. We are usually concerned with the rising, setting and transit times of the moon and planets. Detailed information on other uses is in the explanatory notes that come with the almanac.

(c) Figure 9-1 shows a portion of a recent almanac in case a current one is not available. Note that the MONTH and DATE are given along the left-hand side and the local TIME across the top and bottom. The complex of diagonal and curved lines on the chart represent various events such as sunrise and sunset; and planet's rising, setting and transit times. To find what time a given event happens on a specific date, find the proper event line and trace it to the horizontal line of the date. Then read the time from their intersection point. For example: Saturn transits at midnight (standard time) on September 29.

(d) Determine each of the following:

i. When does Venus rise on September 1?

ii. What time does Saturn set on October 1? What time is sunset on October 1? Will you be able to see Saturn on this evening before it sets? (Hint: Check the time that evening astronomical twilight ends.)

iii. Check the setting time of Mercury on December 22. What do you find? What does this suggest about where one would look for it in the sky?

iv. The best time to view the superior planets is when they are high in the sky near midnight. This is called the "transit" of a planet. When will Saturn be in a good viewing position in the fall of 1996?

v. When will Uranus set on December 1?

DISCUSSION QUESTIONS

1. Do the planets seem to fall in any pattern when plotted on your star chart? If so, what might be the significance of such a pattern?

2. We know that the stars rise and set about 2 hours earlier each successive month. Use the *Sky Gazers Almanac* and get the setting time for Mars or Venus for two dates separated by a month. Do they follow the pattern of the stars? If not, why not?

3. (a) What are the relative advantages and disadvantages of using the charts and other devices?

   (b) What is the major source of error in each device?

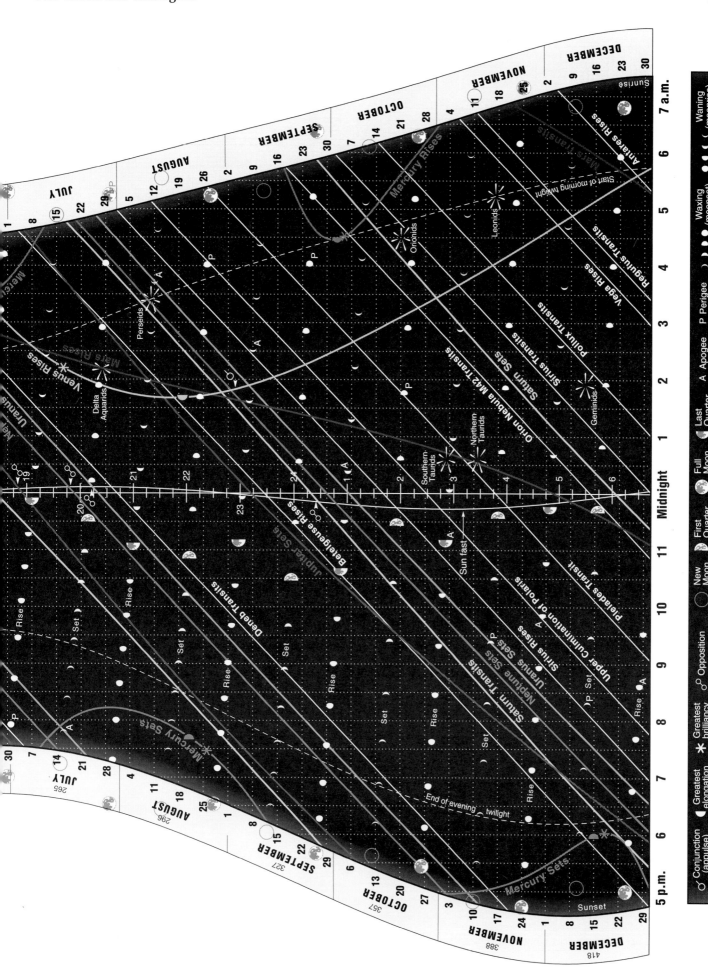

Figure 9-1. A Portion of the Sky Gazer's Almanac 1996.
(From the Sky Gazer's Almanac 1996, with permission of Sky Publishing, Inc.)

PART II

**Basic Astronomy
and the Solar System**

10 Angles and Parallax

PURPOSE AND PROCESSES

The purpose of this exercise is to introduce simple techniques for measuring angles, to use these techniques to become familiar with the relationship between distance, angular size, and linear size; and to study the principles of the parallax method of determining distances to celestial objects. The processes stressed in this exercise include:

Using Numbers
Identifying Variables
Controlling Variables
Using Logic

INTRODUCTION

We use the relationship between the angular size of a known object and its distance almost automatically when we are driving an automobile. A car which "looks half as big" (subtends an angle half as large as another automobile) is approximately twice as far away because automobiles are roughly the same size. We use the fact that we know the approximate linear sizes of common objects such as cars, buildings, and bicycles to estimate their distances from their apparent angular sizes. The mental conversion of an observed angular size into an estimated distance is done easily by most persons for common objects that are relatively close. However, we would have difficulty in estimating the distance to the Goodyear blimp. We don't see it very often and, if it is high in the sky, there is no common object at the same distance to compare it to. The eye can measure only the ANGULAR separations or angular sizes of objects; it takes the mind to convert to real physical measurements. We want to make this process more systematic and quantitative, and use a simple and practical technique for measuring angles to measure distances and/or linear sizes of various objects on the ground.

One of the basic problems in astronomy is determining distances to celestial objects. By definition, *parallax* is one-half of the angle formed at the celestial body by two intersecting lines drawn from the ends of a baseline. In the case of objects in the solar system, the baseline is a diameter of the earth, and in cases of objects outside the solar system the baseline is the diameter of the earth's orbit. Technically these are called *geocentric* (or *horizontal*) parallax and *heliocentric* (or *annual*) parallax, respectively (Figures 10-1 and 10-2).

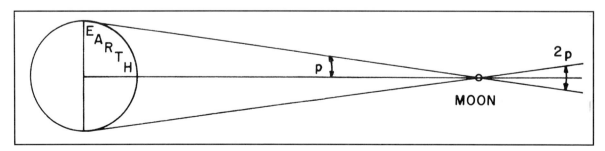

Figure 10-1. Geocentric or Horizontal Parallax.

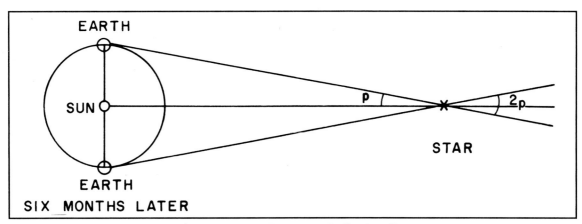

Figure 10-2. Heliocentric Stellar Parallax.

If the moon is observed from the two end points of a diameter of the earth it will be seen in two differ-ent positions with respect to the background stars. The total amount of apparent displacement (2p) is about two degrees, the exact value depending on the earth-moon distance. Since parallax is defined as one-half of this displacement (p), the parallax of the moon is about one degree.

A similar shift is seen for all stars during the course of one year. Such small displacements are involved that the measurement of stellar parallax is most commonly done by photographing a field of stars at in-tervals of six months, and then measuring the relative positions of the stars with a microscope.

Even then, the largest parallax we measure for the nearest star (α Centauri), is less than one second of arc: the size of a quarter (about 1 inch) seen from a distance of about 3.25 miles!

In order to deal with such large distances we have defined a distance unit: if the parallax is one second of arc, the distance is said to be one *parsec*, (a coined word consisting of the first three letters of the words parallax and second). Using the $s = r\,\theta$ equation and a baseline of 1 AU we find that

$$1 \text{ parsec} = 206{,}265 \text{ AU}$$

which is equal to 3.086×10^{13} km. The parsec is a commonly used unit by astronomers since it allows us to use the simple relation

$$d = \frac{1}{p}$$

where   p = parallax in seconds of arc

        d = distance in parsecs.

One parsec is also equal to 3.26 light years.

PROCEDURE

1. MEASURING ANGLES

(a) The first step is to make and calibrate your own scale for measuring angles. A transparent plastic strip marked off in 5° intervals is a simple and convenient scale since it can be carried with you easily. The cali-bration procedure is illustrated in Figure 10-3.

Stand a known distance from a wall (say 15 feet) and hold the plastic strip vertically at arms length be-tween the thumb and index finger of each hand. Line up your right thumb and forefinger with an upper mark on the wall and then your left thumb and fore-finger with the lower mark. If the marks on the wall are 0.087 times the distance-from-the-wall apart (1.305 feet or 15⅔ inches for a distance of 15 feet) your thumbs will be 5° apart. Hold the strip flat on a table and mark the positions of your thumbs on the strip with a permanent magic marker. 10° and 15° intervals can be marked by doubling and tripling the 5° inter-val and 1° intervals can be obtained by subdivisions of a 5° interval.

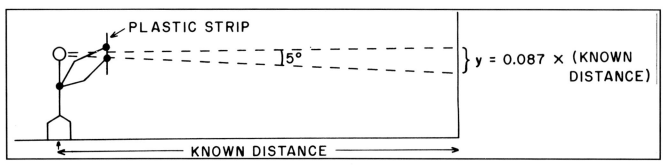

Figure 10-3. Calibration of the Plastic Strip.

(b) Using the plastic strip, measure the angular height of a pole, building, or other object at a number of different distances. Plot your data on a graph and look for a general relationship. Is it linear? If not, can you redefine the variable plotted on one axis and replot the graph so you get a linear relationship?

(c) It makes sense that the larger the distance, the smaller the angle for a given object. This relation can be stated:

$$s = r\,\theta$$

where s = linear size
       r = distance
       θ = angular size

The relation originates from considering a circle (Figure 10-4):

       s = arc length
       r = radius
       θ = angle

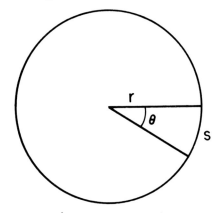

Figure 10-4. s = r θ

and works well for things not on a circle if the angle θ is less than 15°. The only complication is that θ has to be in units of *radians*, where 1 radian is 57°. This comes from the circle too, as for a WHOLE circle

$$s = \text{circumference} = 2\,\pi\,r$$

and

$$\theta = 360°$$

so

$$s = 2\pi\,r = r\,\theta.$$

To work, θ has to be equal to 2π. We get 2π radians equal to 360°, or

$$1 \text{ radian} = \frac{360°}{2\pi} \approx 57°.$$

To sum it all up

$$s = r\,\theta \ (\theta \text{ in radians})$$

or

$$s = r\,\frac{\theta}{57} \ (\theta \text{ in degrees}).$$

Calculate the height of your building or other object in part b in feet or meters (or in "paces" if you paced off the distances).

(d) Determine the height of a classmate at a known distance at the other end of the hall.

(e) Practice using your strip to measure the angles between stars or the moon and stars. Can you calculate distances from these measurements?

2. PARALLAX

Measuring distance using the parallax method involves measuring the angle an object appears to move through compared to a distant background when viewed from two different places. It involves applying the s = r θ relationship with a different pair of "knowns": now we can measure s and θ, and solve for the distance r. A common example of parallax is to hold a finger out in front of you and look at it first with one eye closed, then the other (Figure 10-5). The finger appears to move back and forth even though we know the apparent motion is really caused by looking through each eye in turn. If you measured the parallax angle θ your finger appears to move through, and the distance between your eyes s (often called the *baseline*), you could calculate the distance r to your finger.

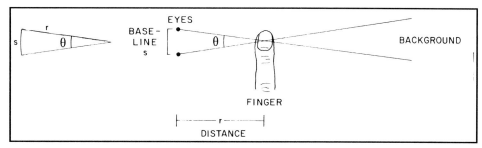

Figure 10-5. A Common Example of Parallax.

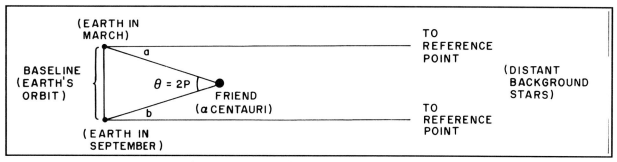

Figure 10-6. Simple Parallax Measurements.

Since

$$s = r\,\theta$$

we can solve for the distance r:

$$r = \frac{s}{\theta}\ (\theta\ \text{in radians})$$

or

$$r = 57\frac{s}{\theta}\ (\theta\ \text{in degrees}).$$

We will use this method to measure the distance of a friend or object down a long hallway or outdoors using a distant object as a reference point.

(a) Place an object or have a friend stand partway down the hall. Mark off a baseline of several feet.

(b) Standing at one end of the baseline, measure the angle between your friend and some reference point as far down the hall as possible (angle a in Figure 10-6). If you can go outdoors, a distant building or radio tower will make an excellent reference object. Repeat the measurement from the other end of the baseline (angle b in Figure 10-6).

(c) Add your two measured angles (a and b) to get the angle θ and find the distance to your friend using s = r θ.

DISCUSSION QUESTIONS

1. List some possible reasons for differences in the observed and measured distances to the "model star" or friend.

2. How does the distance of the "distant" source (representing the background stars) affect parallax measurements?

3. The star Deneb is said to be at a distance of 1600 light years. What should its parallax be? Is it probable that this figure for distance was obtained from parallax measurements?

4. What would be the parallax of the nearest star Alpha Centauri if measurements were made from Mars? Refer to your text to obtain the distance to Alpha Centauri.

11

Kirchhoff's Laws and Spectroscopy

PURPOSE AND PROCESSES

The purpose of this exercise is to become familiar with Kirchhoff's Laws and the techniques of laboratory spectroscopy. The processes stressed in this exercise include:

Observing
Formulation of Models
Inferring

INTRODUCTION

It is important to understand the properties of light because most of an astronomer's information about celestial objects arrives in the form of electromagnetic waves. Astronomers, both in the laboratory and at the telescope, have spent a great deal of time in devising techniques to decode the message of light in order to yield as much information as possible. By examining a mere pinpoint of light we can now determine many of the physical properties of the body from which the light originates: properties such as temperature and chemical composition.

One of the astronomer's most useful tools is the spectroscope or spectrograph, which spreads light into its spectrum or rainbow of composite colors. We find that many properties of light can be explained in terms of its behavior as an energy wave. The *wavelength* of light determines its color: red light is made up of waves almost twice as long as those of violet light. The various colors of light we see are simply waves of different lengths. The order of colors is the same as that seen in a rainbow: red (longer waves) through orange, yellow, green, blue, and violet (shorter waves).

In fact, light (or as it is more generally called *electromagnetic radiation*) comes in many colors we can't see: for example, ultraviolet rays are those just shorter than violet, and infrared rays are those just longer than red (Figure 11-1). Our visible light is just a small part of the whole electromagnetic spectrum, which includes even x-rays (with very short wavelengths) to the longest radio waves. The white light we see is just a combination of all the colors of visible light.

An instrument used to form a spectrum from a light source is called a *spectroscope* if the spectrum is viewed visually, or a *spectrograph* if it is recorded on photographic film or plates. The light in either instrument is dispersed or spread out into its colors by a prism or diffraction grating. A slit to make a thin beam of light and one or more lenses are also necessary. A prism spectrograph is schematically represented in Figure 11-2.

A prism works by bending the different colors of white light by different amounts as they pass through it. A grating is made by ruling a series of very fine parallel lines (often 20,000-30,000 lines per inch) on a mirror or piece of glass. It produces a spectrum by the more complicated processes of diffraction and interference of the light. The grating has an advantage over the prism in that it has equal dispersion at all wavelengths, while prisms disperse blue light more than red. Replica gratings can now be produced at very low cost.

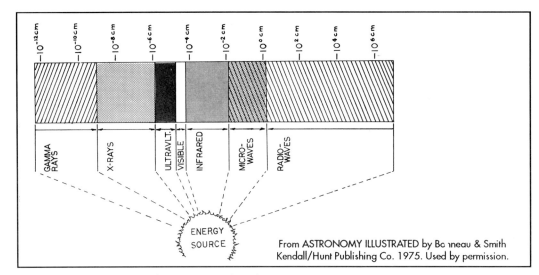

Figure 11-1. The Electromagnetic Spectrum.

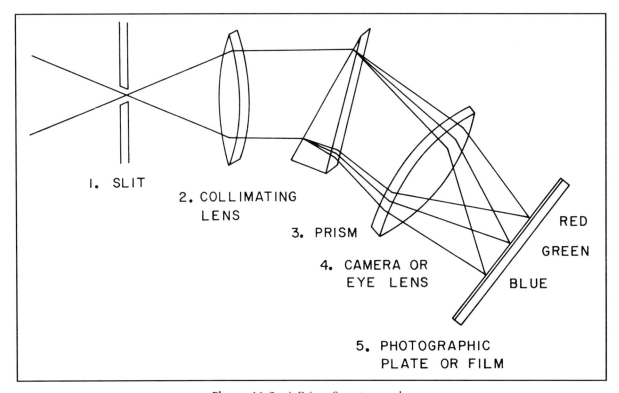

Figure 11-2. A Prism Spectrograph.
1. The slit is placed at the focal point of the objective.
2. The collimating lens makes the light parallel.
3. The prism disperses the light.
4. The camera lens focuses the light on photographic film.
5. The film records the spectrum as an image of the slit for each individual color or wavelength.

In the 1800s when spectra from various luminous objects were first being studied it was found that all spectra were not like our customary rainbow: some have bands of color missing, and others have bright bands of color. Spectra were divided into three main types by Gustav Kirchhoff when he proposed explanations for their origins. These explanations are Kirchhoff's Three Laws of Spectral Analysis.

FIRST LAW:

A luminous solid, liquid or very dense gas emits light of all wavelengths, thus producing a *continuous spectrum* or *continuum*. This spectrum appears as a continuous change of colors from red to a deep violet (like a rainbow). We should remember that it really extends beyond the red through the infrared and beyond the violet into the ultraviolet regions even though our eyes cannot detect such wavelengths. The continuous spectrum is nearly the same for all substances and varies in total brightness with the temperature of the solid or liquid. The distribution of intensity as a function of color also varies slightly as the temperature is changed. Continuous spectra give almost no indication of chemical composition.

SECOND LAW:

A rarefied or low-pressure gas emits light whose spectrum contains bright lines, sometimes superimposed on a faint continuous spectrum. This is a *bright line* or *emission spectrum*. It appears as a set of distinct sharp lines of different colors, separated by darker spaces. The wavelengths of specific lines present are characteristic of the atoms of the gas. Each type of atom or element in the periodic table has its own unique set of lines. If the gas is a combination of several types of atoms then the spectrum will contain lines characteristic of each element. The emission spectrum therefore is of great importance in determining the chemical composition of the gas.

THIRD LAW:

If a continuous spectrum from a luminous source passes through a cool tenuous gas, certain wavelength regions will be extracted from the continuum causing dark lines (or absorption lines) to appear. This *absorption spectrum* looks like a continuum with certain sharp lines blotted out. It is found that these lines are at exactly the same positions that would be occupied if the same gas were emitting rather than absorbing. Thus these absorption lines are just as useful as emission lines in identifying the composition of the gas.

The three types of spectra are depicted in Figure 11-3.

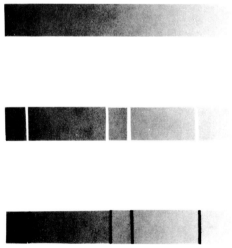

Figure 11-3. Three Types of Spectra.

PROCEDURE

1. IDENTIFICATION OF ELEMENTS FROM THEIR SPECTRA

It is possible to produce emission spectra of many elements by heating a wire dipped in certain chemical salts, or by using gas discharge tubes. Produce as many spectra as possible and view each with a spectroscope. (See data sheet on page 00.)

For each:

    i. Record the general color of the flame or tube as it appears visually.

    ii. Sketch the spectrum indicating both the COLORS and RELATIVE SPACING of the lines. Colored pencils may be used if available.

    iii. If the element being observed is unknown, sketch it as above, and identify it by comparison with other spectra you have recorded or with a wall chart of common spectra.

(a) Gas Discharge Tubes

Each tube has been evacuated and then filled with a certain type of gas. When a high voltage is applied across the tube the gas glows and emits light.

BE CAREFUL IF IT IS NECESSARY TO REMOVE THE GAS TUBES FROM THE TUBE HOLDERS. The use of asbestos gloves is often required.

(b) Flame Test

You will be supplied with a platinum test wire and a bunsen burner and several salts such as Calcium Oxide (CaO), Strontium Bromide (SrBr), and Potassium Bromide (KBr).

It is possible to produce a gas from each of the salts (Ca gas, Sr gas, or K gas) for a short time (several seconds at most) by dipping the platinum wire into the salt and holding it over the bunsen burner flame. Be sure to dip the wire into hydrochloric acid before dipping into the salt each time. Since the duration of the colored flame is short, it may be helpful to have one person produce the flame while several others view it.

If a propane torch is used instead of a bunsen burner, observe and record the appearance of its spectrum. Note the wide absorption bands and indicate their approximate positions in your drawing. These absorption bands are called the Swan spectrum and are due to carbon (C_2) in the flame of the torch. They are of special interest to the astronomer as they are often observed (and thus imply the presence of C_2) in the spectra of comets.

2. COMMON LIGHT SOURCES

(a) Use the spectroscope to examine the spectra produced by as many of the following common light sources as possible:

    An incandescent light bulb
    Fluorescent lights
    Street lights and spotlights of various types
    Matches
    A candle
    A cigarette lighter
    A firefly.

Indicate the type of spectrum each produces (continuous, absorption or emission). If it has lines sketch them as you did above and identify the elements present.

(b) Look at the light from an incandescent source with and without a holmium oxide or similar filter. Note what changes the filter makes in the spectrum's appearance. Use two filters, and again note any change.

(c) Observe the solar spectrum and name its general type. Sketch some of the more prominent lines and identify them if possible. (Use care not to look directly at the sun!)

11. Kirchhoff's Laws and Spectroscopy

Spectra of Known Gases

Gas	Spectrum	General Color
1. _____		_____
2. _____		_____
3. _____		_____
4. _____		_____
5. _____		_____
6. _____		_____
7. _____		_____
8. _____		_____
9. _____		_____
10. _____		_____

Spectra of Unknown Gases and Their Identifications

Gas	Spectrum	General Color
1. _____		_____
2. _____		_____
3. _____		_____
4. _____		_____

DISCUSSION QUESTIONS

1. A faint continuum is often observed along with the bright emission lines from a gas discharge tube. Can you suggest its origin?

2. If an "air" tube is available, compare its spectrum to those of the other gases you observed. Do you note any similarities? Discuss.

3. If spectra were viewed with filters in part 2 above, what differences did one or two filters make in the spectrum? Can you generalize from your observations?

4. What does Kirchhoff's explanation of the type of spectrum of the sun tell us about the sun's physical nature?

Image Size—Focal Length Relationship

The purpose of this exercise is to determine the relationship between the size of the image of an extended object and the focal length of the optical system producing it. The processes stressed in this exercise include:

Using Numbers
Interpreting Data
Controlling Variables
Inferring
Prediction

INTRODUCTION

The size of an image of a distant object depends only on the focal length of the mirror or lens system producing it. The *image scale* of a telescope is a constant for a particular optical system. It gives the size of the image for each degree of arc the object subtends, often expressed in units of inches per degree, or in millimeters per arc minute.

We wish to find the general relationship between focal length and image size for a one degree object. Our data consists of several photographs of the moon taken with cameras of various focal lengths. The values are given in Table 12-I.

Table 12-I

Series Number	Focal Length (mm)
1	100
2	205
3	1500
4	3048
5	4570

Whenever we wish to find a general relationship, we are looking for a pattern in the data. One of the best ways to find such a regularity is to use a graph. If the data fall along a line passing through the origin, our task is relatively simple, as the relationship may then be expressed in terms of an equation of the form:

$$y = mx$$

or

$$s = mf$$

where s = image size
f = focal length
m = slope of the line
= $(s_2 - s_1)/(f_2 - f_1)$ for any two points on the line.

PROCEDURE

1. Measure the size (diameter) of each image in figure 12-1 to a tenth of a millimeter. Your accuracy will be improved if you make several measurements for each photograph and calculate an average value for each series. For series number five, you will have to measure the size of a crater and compare it to the same crater in series four. Set up a proportion where the ratio of crater sizes is equal to the ratio of image sizes.

2. Enter your data in a table, allowing columns for each step in your calculations. Remember that:

   (a) The prints are enlarged two times, and

   (b) The angular size of the moon is approximately ½ degree.

3. Plot the image scale (the image size for an object of one degree of arc) against the focal length for each of the five series of photographs (labeled graph paper is given in Appendix 4). Fit the best straight line through your data, trying to equalize deviations of the points on either side of the line. Include the point for $f = s = 0$.

4. Determine the slope of your line and express the image size—focal length relationship in a linear equation.

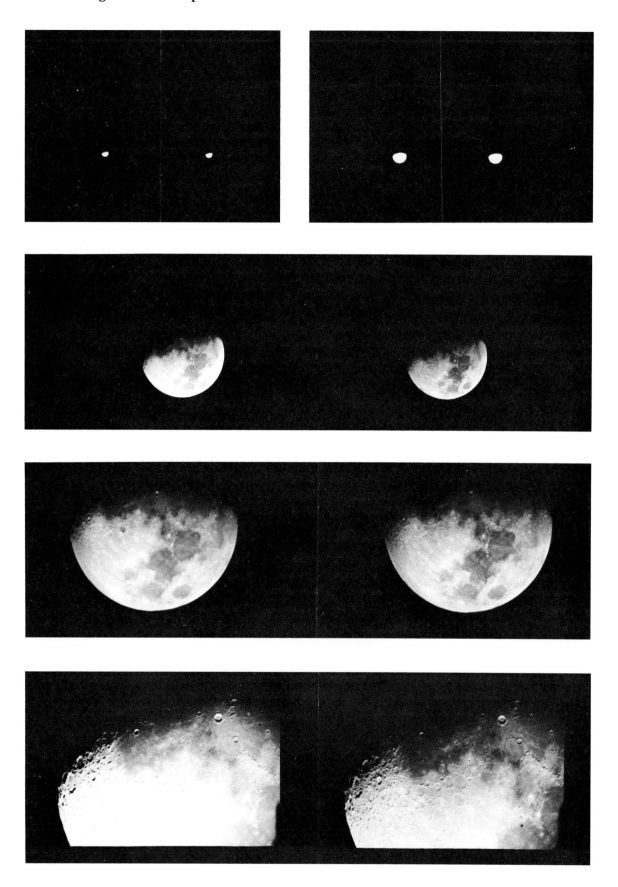

Figure 12-1. Photographs of the Moon Taken with Various Lens Systems.
(Courtesy William A. Lane, University of Iowa Observatory)

12. Image Size—Focal Length Relationship

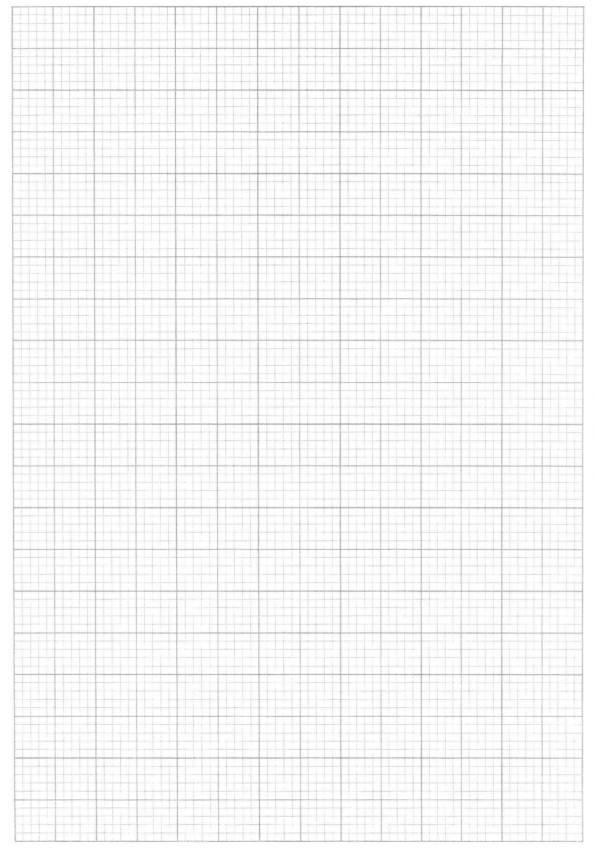

Image Size for a 1° Object (mm)

Focal Length (mm)

DISCUSSION QUESTIONS

1. Which of your data points is least accurate and why? How should this be accounted for in fitting a line to the data?

2. What would the image size of the moon be if you photographed it at Cassegrain focus of a 24-inch telescope with a focal length of 9600 mm?

3. An astronomer wants to determine whether or not he or she can photograph Castor and Pollux at the same time on a 5 × 8-inch plate using a lens system with a 1200 mm focal length. If the two stars are approximately 5 degrees apart, will it be possible to make the photograph?

4. Predict the size of the image of Jupiter if photographed at opposition with a lens having a focal length of 9600 mm. (Assume Jupiter is a disk of 1.43×10^5 km and is 6.29×10^8 km from the earth.)

13 | Length of the Sidereal Day

PURPOSE AND PROCESSES

The purpose of this exercise is to determine the length of the sidereal day from a photograph of the circumpolar region of the sky. The processes stressed in this exercise include:

Using Numbers
Designing Experiments
Using Logic
Inferring

INTRODUCTION

The length of the *sidereal day* can be defined as the time interval between two successive meridian transits of the Vernal Equinox. It is time based on the earth's rotation with respect to the celestial sphere or stars rather than with respect to the sun (as for solar time). In order to measure the length of the sidereal day we must measure accurately the apparent motion of the stars around the sky. Since this is difficult to do for a full day, it is convenient to record the motion of the stars on photographic film for a shorter, but well-known time duration.

The area of the sky around the north celestial pole (near Polaris) works particularly well for this purpose. The stars in this area are called *circumpolar* because they appear to circle the pole rather than rise or set.

PROCEDURE

1. If possible, take a photograph of circumpolar star trails. Attach your camera to a tripod or clamp and center the field of view near Polaris. Open the shutter for a period of 10 minutes to several hours, and record your exposure time accurately. High sensitivity film such as Tri-X is best. Do not move your camera during the exposure.

2. If you are unable to obtain a photograph, Figure 13-1 may be used. The shutter was opened for 15 minutes, the lens cap replaced for 5 minutes, and removed again for 90 minutes to determine the direction of rotation. (These measurements were made with a stopwatch.)

3. Measure the star trail arc lengths and devise a method for determining the length of the sidereal day. The brightest stars in Ursa Minor are labeled in Figure 13-1. The coordinates of these stars are given in Table 13-I. Outline your method in your lab report, and make all necessary measurements and calculations. (A sheet of polar-coordinate tracing paper might make measurements easier.)

Table 13-I

Star	RA (2000)		dec (2000)	
β UMi	14	50.7	+74	09´
γ UMi	15	20.7	+71	50´
δ UMi	17	32.2	+86	35´
ε UMi	16	46.0	+82	02´

Figure 13-1. Circumpolar Star Trails.
(Courtesy Dean Ketelsen, University of Iowa)

DISCUSSION QUESTIONS

1. Does your measured length of the sidereal day equal the length of the solar day? Would you expect the two to be equal? Why or why not?

2. What is the rotational velocity of the earth at the equator? At your latitude?

14 | Determining the Mass of the Moon

PURPOSE AND PROCESSES

The purpose of this exercise is to determine the mass of the Earth's moon using data from a lunar orbiting satellite. The processes stressed in this exercise include:

Using Numbers
Interpreting Data
Using Logic
Questioning
Inferring

INTRODUCTION

Working with data obtained by Tycho Brahe, Johannes Kepler determined Three Laws of Planetary Motion:

1. The orbits of the planets are ellipses with the sun at one focus.

2. A line segment connecting the sun and a given planet sweeps out equal areas in equal time intervals.

3. $P^2 = a^3$ where "P" is the sidereal period of the planet in years and "a" is the semimajor axis of its orbit in astronomical units (A.U.). (The period describes the time of one complete revolution about the sun and the semimajor axis is defined as shown in Figure 14-1.)

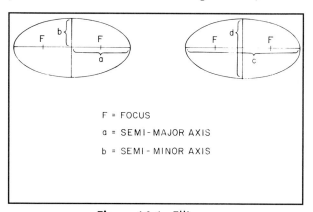

F = FOCUS

a = SEMI-MAJOR AXIS

b = SEMI-MINOR AXIS

Figure 14-1. Ellipses.

Newton later generalized these laws to apply to any two bodies in orbital motion about each other and the third law reads:

$$(m + M)\, P^2 = \frac{4\pi^2}{G}\, a^3$$

where m and M are the two masses
   P is the period of mutual revolution
   G is the universal gravitational constant
   a is the semimajor axis of their relative orbits.

This form reduces to the one given by Kepler when we consider that a planet's mass is negligible compared to that of the sun

$$(m_{planet} + m_{sun} \approx m_{sun}):$$

$$m_{sun} P^2 = \frac{4\pi^2}{G}\, a^3$$

If m is expressed in solar masses, P in years, and a in astronomical units, G will take on such a value that $4\pi^2/G$ is equal to one. So we have

$$m_{sun} P^2 = 1\ P^2 = a^3$$

which agrees with Kepler's Law.

PROCEDURE

Explorer 35 was launched from Cape Kennedy on July 19, 1967, and placed in orbit about the moon on July 22. The 230-pound NASA spacecraft carried instruments for measuring solar x-rays and energetic particles; the solar wind in interplanetary space; and the magnetic properties of the moon as well as the interaction of the solar wind with the moon. The mission objectives were accomplished and the spacecraft continued to operate until June 1973. Table 14-I gives a sample set of positional data of Explorer 35 in its elliptical orbit about the moon.[1] The time interval between entries is 15 minutes, the unit of length is the radius of the moon, and the center of the coordinate system is the center of the moon.

1. Plot the data and find the major axis and foci. Verify Kepler's First Law. (Hint: refer to the definition of an ellipse or the "string method" for drawing one.) (Use graph paper on page 00.)

2. Find the semimajor axis, minor axis, semiminor axis, the eccentricity, and the period of orbit. (The eccentricity is an index of the "flatness" of the ellipse, and is defined as the distance between the foci divided by the length of the major axis.)

3. Show that Kepler's Second Law is valid.

4. Using Kepler's Third Law (Newtonian form) find the mass of the moon in kilograms. Some additional information:

radius of the moon = 1738 km
$$G = 6.668 \times 10^{-8}\ cm^3/gm\ sec^2$$
$$= 8.642 \times 10^{13}\ km^3/kg\ hr^2$$
so that $4\pi^2/G = 4.568 \times 10^{13}\ kg\ hr^2/km^3$

---

1. Courtesy Goddard Space Flight Center/National Aeronautics and Space Administration.

Table 14-I

Elapsed Time	X (Lunar Radii)	Y (Lunar Radii)	Elapsed Time	X (Lunar Radii)	Y (Lunar Radii)
$0^h\ 00^m$	–3.62	1.04	$6^h\ 00^m$	–0.27	4.86
$0^h\ 15^m$	–3.46	0.63	$6^h\ 15^m$	–0.56	4.95
$0^h\ 30^m$	–3.25	0.20	$6^h\ 30^m$	–0.84	5.01
$0^h\ 45^m$	–2.97	–0.22	$6^h\ 45^m$	–1.12	5.03
$1^h\ 00^m$	–2.60	–0.65	$7^h\ 00^m$	–1.38	5.04
$1^h\ 15^m$	–2.14	–1.03	$7^h\ 15^m$	–1.64	5.00
$1^h\ 30^m$	–1.55	–1.37	$7^h\ 30^m$	–1.89	4.95
$1^h\ 45^m$	–0.85	–1.58	$7^h\ 45^m$	–2.14	4.87
$2^h\ 00^m$	–0.03	–1.59	$8^h\ 00^m$	–2.37	4.77
$2^h\ 15^m$	+0.78	–1.32	$8^h\ 15^m$	–2.59	4.65
$2^h\ 30^m$	1.45	–0.79	$8^h\ 30^m$	–2.80	4.50
$2^h\ 45^m$	1.87	–0.11	$8^h\ 45^m$	–2.99	4.33
$3^h\ 00^m$	2.09	+0.58	$9^h\ 00^m$	–3.17	4.14
$3^h\ 15^m$	2.16	1.22	$9^h\ 15^m$	–3.33	3.93
$3^h\ 30^m$	2.11	1.82	$9^h\ 30^m$	–3.49	3.69
$3^h\ 45^m$	1.99	2.35	$9^h\ 45^m$	–3.59	3.42
$4^h\ 00^m$	1.82	2.81	$10^h\ 00^m$	–3.69	3.15
$4^h\ 15^m$	1.61	3.22	$10^h\ 15^m$	–3.77	2.85
$4^h\ 30^m$	1.37	3.59	$10^h\ 30^m$	–3.81	2.52
$4^h\ 45^m$	1.11	3.90	$10^h\ 45^m$	–3.83	2.20
$5^h\ 00^m$	0.85	4.16	$11^h\ 00^m$	–3.81	1.83
$5^h\ 15^m$	0.58	4.40	$11^h\ 15^m$	–3.76	1.46
$5^h\ 30^m$	0.28	4.58	$11^h\ 30^m$	–3.65	1.06
$5^h\ 45^m$	0.00	4.74	$11^h\ 45^m$	–3.51	0.65

*Adapted from data from James A. Van Allen, The University of Iowa.
(Courtesy Goddard Space Flight Center/National Aeronautics and Space Administration)

14. Determining the Mass of the Moon

Exercise 14.   Explorer 35 and Kepler's Laws

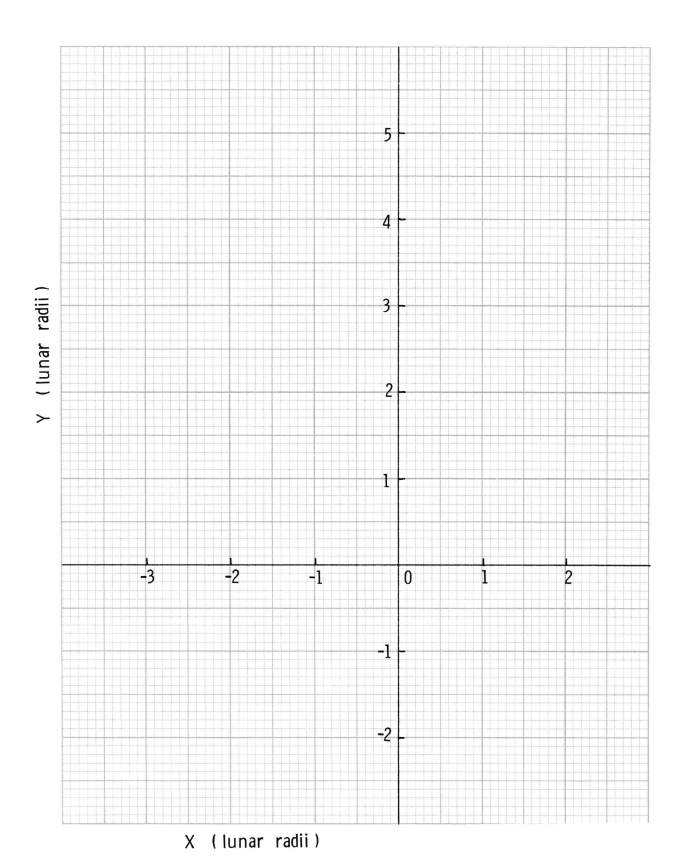

X   (lunar radii)

DISCUSSION QUESTIONS

1. Sketch an ellipse with an eccentricity of 1; of 0.

2. Discuss any variations in the measurements taken to verify Kepler's First and Second Laws. What were the uncertainties in your measurements? How do they compare to the variations in your data?

3. Estimate your uncertainties in determining the mass of the moon. What has a greater effect, uncertainties in "a" or those in "P"? Why?

15 Lunar Features and Mountain Heights

PURPOSE AND PROCESSES

The purpose of this exercise is to obtain photographs of the lunar surface, study the different types of lunar features, and calculate the heights of various lunar mountains or crater walls. The processes stressed in this exercise include:

Observing
Using Numbers
Controlling Variables
Using Logic
Interpreting Data

INTRODUCTION

The moon has a great number of surface features available for observation. The large maria are visible with the naked eye and craters are easily seen with binoculars. With the greater magnification of telescopes and satellite photographs, many other types of surface features can be identified. The shadows cast by mountain peaks and crater walls are useful in mapping the three-dimensional lunar topography. The height of such features can be calculated by triangulation after the shadow length and the local altitude of the sun have been determined.

PROCEDURE

1. TYPES OF SURFACE FEATURES

(a) Photograph and print a series of pictures of the moon near first or third quarter, with as high a magnification as is practical. (You might want to refer to Exercise 6 "Astronomical Photography.") If it is not possible to take your own pictures, use the one accompanying this exercise. Lunar orbiter photographs can also be used if they are available. (See Appendix 3 "Equipment Notes.")

(b) Examine closely the photographs of the moon's surface. Sketch the large-scale features and identify by name the major maria and craters. List and describe as many different types of features as you can find, name one of each, and give its approximate coordinates.

2. SIZE OF LUNAR FEATURES

One can determine lunar feature sizes rather easily by measuring feature sizes on the photograph (with either a millimeter ruler or a micrometer eyepiece) and using simple proportion.

(a) Measure the diameter of your lunar image from your print.

(b) Select five lunar features (maria or prominent craters) and measure their diameters to the nearest tenth millimeter.

(c) The accepted lunar diameter is 3476 kilometers. Using this value, set up a proportion:

$$\frac{MFD}{AFD} = \frac{ILD}{ALD}$$

where:

MFD = your measured feature diameter (mm)
AFD = actual feature diameter (km)
ILD = image lunar diameter (mm)
ALD = actual lunar diameter (3476 km)

(d) Solve for the actual feature diameters in kilometers.

3. HEIGHT OF A MOUNTAIN OR CRATER WALL USING A LUNAR PHOTOGRAPH

Determining lunar feature heights is basically a two-step process. First the *scale height* of the feature as it appears in the scale of the photograph is found. This scale height can then be converted the real feature's height on the moon knowing the radius of the moon and its apparent radius in the photograph.

(a) In Figure 15-1 let M be a surface feature whose height is to be measured. The *terminator* is the sunset or sunrise line on the surface of the moon. It is not a well defined line as an examination of Figure 15-4 will show. However, it is necessary to estimate its location to determine lunar feature heights.

(b) To more easily understand this method of measuring scale heights, visualize rotating the moon upward so that M is on the top edge as in Figure 15-2. In the enlarged Figure 15-3 we can see that triangles TOM and MAP are similar: AP and TM are parallel lines and angle APM is equal to angle TMO. Therefore, we can set up ratios of corresponding sides

$$\frac{TM}{OM} = \frac{MP}{AP}$$

where TM represents the distance of the feature from the terminator

OM represents the radius of the moon
MP represents the scale height of the feature
AP represents the length of the feature's shadow.

Rearranging to find the feature height we have

$$MP = AP \; \frac{TM}{OM}$$

(c) Measure AP, TM and OM to the nearest tenth millimeter for several features on your lunar photograph. The photograph in Figure 15-4 has a lunar radius of 267 mm. Calculate the scale heights of the features you selected.

(d) Measured scale heights (in mm) may be converted to real heights of features (in km) on the moon using the following proportion:

$$\frac{\text{scale height of feature (mm)}}{\text{real height of feature (km)}} = \frac{\text{lunar radius in photograph (mm)}}{\text{real lunar radius (km)}}$$

The lunar radius is 1738 km. Calculate the heights of your features in km.

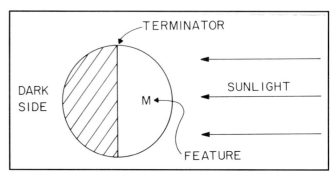

Figure 15-1. Lunar Surface Feature.

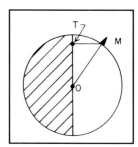

Figure 15-2. Visualized Upward Rotation of the Moon.

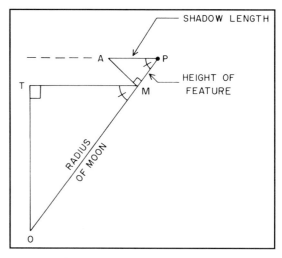

Figure 15-3. Similar Triangles.

Figure 15-4. The Moon Near First Quarter.
(Courtesy Larry A. Kelsey, University of Iowa Observatory)

16

The Moon's Sidereal Period

PURPOSE AND PROCESSES

The purpose of this exercise is to determine the sidereal period of the moon from a series of pictures showing its motion with respect to the planet Venus. The processes stressed in this exercise include:

Using Numbers
Interpreting Data
Identifying Variables
Controlling Variables
Inferring

INTRODUCTION

The moon passes near a bright planet occasionally, providing a good opportunity to observe and photograph its angular velocity. It is easy even for beginners to take photographs from which direct measurements can be made. The sidereal period of the moon can then be calculated.

PROCEDURE

1. MOON'S ANGULAR VELOCITY

If possible try to take your own pictures of the moon as it passes near a bright planet. The accompanying photographs (Figure 16-3) were taken of the crescent moon and the planet Venus on April 16, 1972, from Cedar Falls, Iowa. They were taken with a 135 mm lens and a single lens reflex camera on Tri-X film at f/3.5 and 1/30 second exposure time. The four pictures were taken at 7:30 P.M., 8:00 P.M., 9:00 P.M., and 10:00 P.M., respectively.

A visual examination shows the high angular velocity of the moon relative to the planet. The photos were taken facing west, and the motion of the moon relative to Venus is eastwardly. Since Venus was nearly at greatest eastern elongation it had little tangential or crosswise motion relative to the earth, and most of its motion was radial or along our line of sight (Figure 16-1). Most of the observed motion is therefore due to the motion of the moon. However, a small correction will be made later for the motion of the earth.

(a) Using a sheet of tracing paper, carefully trace the outline of the moon from the first picture of Figure 16-3, and mark the position of Venus. Move the tracing paper to each successive photograph, center your moon tracing on the moon's image, and mark Venus' position for each. (You may need to rotate your drawing slightly to align the moon accurately.) Your data should produce a sketch similar to Figure 16-2. Even though it is the moon that is moving, measurements are more accurate when that motion is transferred to Venus. The image of the moon is larger providing a clearer way to orient the sketch. Since Venus is nearly a point, it is easier to use it to trace out the straight line path of its relative motion.

(b) Determine a plate scale to translate linear measurement from the photograph into angular measurements in the sky. Using a compass, trace out a full circle that best fits the circumference of the moon. The

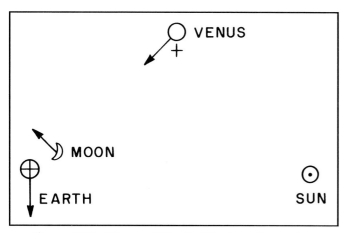

Figure 16-1. Venus at Greatest Eastern Elongation.

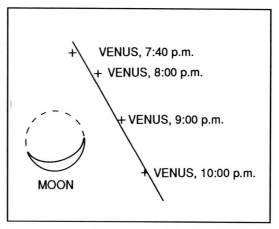

Figure 16-2. Relative Motion of the Moon and Venus.

angular diameter of the moon for this date was 31 minutes of arc. Using your measured diameter of the image in millimeters, calculate a plate scale expressed in minutes of arc per millimeter.

(c) Measure the distance the moon traveled in each time period on your traced diagram, and convert your values to minutes of arc. Determine the angular velocity of the moon for each time period. (Note that the motion of the moon from point A to point B was only 20 minutes.) Average your velocity values to determine the average velocity of the moon (ω) in arc minutes per hour.

2. CORRECTION FOR EARTH'S MOTION

Refer back to Figure 16-1, showing the relative positions of the moon, Earth and Venus (not drawn to scale). Arrows indicate the directions of motion of the earth and moon. Even though the moon is moving with the earth in its orbit, the angle formed between Earth, Venus and the moon is increasing because of the earth's orbital velocity. This angular velocity (ω_E) relative to Venus is roughly that of the earth relative to the sun which equals $360°$/year or about $0.985°$/day. (This value can also be expressed as $2.46'$/hr.) Add the angular velocity of the Earth to the measured velocity of the moon (ω) to give the velocity of the moon relative to a nonmoving earth (ω_o):

$$\omega_o = \omega + \omega_E$$

where ω_o = velocity of the moon with respect to the earth

ω = measured velocity of the moon with respect to Venus

ω_E = velocity of Venus with respect to the earth.

3. THE MOON'S SIDEREAL PERIOD

The entire orbit of the moon is $360°$ or $21,600'$. Set up a proportion to find the sidereal period of the moon in hours and in days.

Figure 16-3a. 7:40 P.M.

Figure 16-3b. 8:00 P.M.

Figure 16-3c. 9:00 P.M.

Figure 16-3d. 10:00 P.M.
(Courtesy: Darrel Hoff, University of Northern Iowa)

DISCUSSION QUESTIONS

1. What are some of the possible sources of error in your work? Estimate their magnitudes. How much will such errors affect your value for the moon's sidereal period? Explain.

2. How would not correcting for the earth's orbital motion in part 2 of the Procedure affect your result? Calculate the orbital period of the moon WITHOUT this correction, and compare your result with the value of the synodic period of the moon. Discuss.

17

The Moon's Geologic History

PURPOSE AND PROCESSES

The purpose of this exercise is to assemble a photographic mosaic of the moon's surface and study the relative geologic ages of the lunar surface features. The processes stressed in this exercise include:

Observing
Interpreting Data
Using Logic
Inferring

INTRODUCTION

Prior to the space age people could only speculate about the surface of the moon. Common ideas in the 1950s were that the surface would be like volcanic slag or a frothy rock with a consistency similar to crunchy snow. Some scientists believed that the continual impact of meteorites had turned the maria into large basins filled with thick layers of fine volcanic dust. Were this the case, lunar landers might sink deep into these seas of dust.

In 1965, prior to the Apollo lunar landings, extensive examination of the surface was begun using lunar orbiter satellites which sent high-resolution images of the lunar surface back to the Earth. The entire lunar surface appeared to be covered with a layer of pulverized ejecta caused by the repeated impacts of meteorites. This ejecta varied in size from large boulders to finely ground dust.

Lunar surface features were formed by two primary processes: meteoric impact and volcanism. Radiometric dating has allowed us to reconstruct major events in the lava flow and meteoric bombardment history. The meteoric impact rate was highest in the first few hundred million years. In the 1960s the lunar time scale was broken down into three parts: pre-mare time (when the impact rate was very high), the mare-forming interval (when large amounts of lava flooded the surface), and post-mare time (which has been relatively quiet, with sporadic meteorite impacts).

Some lunar surface features can be explained in terms of a combination of impact and volcanic processes. For example, many of the moon's craters presumed to have been formed by meteorite impacts have central peaks. Several mechanisms have been proposed to explain these peaks, including 1) Volcanic extrusions triggered by the impact; 2) "Slumping" of materials from the crater rim into the center; 3) Slow rebound of material from underneath the crater pushing up and leading to volcanism in the center; and 4) Rebound immediately after the impact (similar to the rebound seen in slow-motion photographs of a drop of water falling into a puddle). C. A. Wood showed in 1968 that central peaks are most common in the youngest and largest lunar craters.

Secondary impact craters also are believed to have been produced by meteorites falling on the lunar surface. Fragments from the initial impact, as well as fragments from volcanic eruptions, form these secondary impact craters. A number of different types of craters can be formed from different processes (including primary and secondary impacts, volcanic, and collapse events) on a lunar or planetary surface.

The spatial relationships and "overlaps" of the different craters, as well as other features, can be used to estimate relative ages of lunar or planetary surface features. This technique works particularly well for the moon, which shows minimum effects of weathering.

Geologic information can be used to interpret the history of the lunar surface using high-resolution photographs and knowledge of the geologic processes that are likely to have acted on the surface. One tool is identifying different rock units (lava flows, ejecta from craters, overlaying ray material, etc.) to determine the relative ages of the features. In general, unless the surface has been disturbed by weathering or geologic activity, the oldest features will be overlain by the youngest. You can construct an imaginary cross-section of the layers, showing the relative positions (and ages) of the various features. On the moon, for example, you can see that parts of the ancient upland surfaces have been covered by bright rays of ejecta from younger craters.

Relative ages of lunar features can be estimated on the basis of three simple principles:

1. **Superposition:** Younger formations sit on top of older formations. For example, small craters inside of large craters always are younger than the large craters in which they sit. Likewise, impact basins or craters filled with lava flows must be older than the flows that filled them.

2. **Crosscutting:** Younger features always cut across older features. For example, on the moon, the Alpine Valley cuts across the Alps and therefore is younger. Bright rays always are superposed on other features and therefore are younger.

3. **Crater counts:** Old areas of the lunar surface are covered with numerous craters, while younger areas are covered with fewer craters. In addition, the very oldest regions of the lunar surface have many large craters.

We can use this kind of reasoning to estimate a sequence of events on the lunar surface. After the sequence of relative ages has been determined, we can address absolute ages. The most accurate ages are derived from radioisotope dating of rock samples collected from the moon during various Apollo and Luna missions. The absolute dates of some prominent lunar formations are shown in Table 17-I. Note that features without a stated date are listed in the best-known relative (chronological) order.

PROCEDURE

1. If possible, obtain your own high-magnification photographs of the lunar surface to use in making a lunar mosaic. See Exercise 6. "Astronomical Photography" for hints for taking your own pictures.

2. Figures 17-1 through 17-8 are high-magnification lunar photographs taken by an astronomy student with the Central Arizona College 24-inch telescope. T-MAX 400 film (35-mm format) was used at the f/13 Cassegrain focus. Remove these pages, trim the edges, and fit them together (using transparent tape) to form a composite mosaic of the lunar surface. Carefully match areas of similar gray tones when putting the photographs together.

3. Working alone or in small groups, study the assembled lunar mosaic and the full-moon photograph in Figure 17-9. (Figure 17-9 was taken on T-MAX 400 4×5-inch sheet film with the same Central Arizona College telescope, at f/13, with a 1/500 second exposure time.) Note that the full-moon photograph shows rayed craters better than Figures 17-1 through

17-8 taken at a higher magnification. Figures 17-10 and 17-11 are provided to allow you to identify major lunar features. Determine the relative ages of the features in Table 17-II (listed in alphabetical order), and briefly state your reasons for placing each where you did in relative chronological order.

4. Make a chronology of the evolution of the lunar surface using your relative ages (determined above) and the radiometric ages listed in Table 17-I. Number the oldest feature "1" so that the youngest features have the highest numbers.

5. Divide your chronology into several (four or five) eras of geologic activity and give them descriptive names. Consult your text, reference 1, or your instructor to compare your names to those used by lunar geologists. (Eras can be named after the most prominent features formed during the time period they represent.)

6. Label the features listed in your chronology on your lunar mosaic and number them according to their chronological order (as determined in part 4 above). Use a different colored marker to label the features assigned to each different era. Discuss any patterns that you can see.

7. Identify and list several additional lunar features belonging to each of the eras you have created.

Table 17-I
Absolute Dates of Some Prominent Lunar Features

Formation	Radiometric Age (b.y.)	Estimated Age (b.y.)
Tycho		.100
Copernicus		.900
Eratosthenes		2.0 ?
Oceanus Procellarun Lava Flow	3.29–3.38	
Eratosthenes Basalts	3.2	
Mare Crisium Lava Flows	3.3	
Mare Imbrium Lava Flows	3.3–3.7	
Mare Fecunditatis Lava Flows	3.41	
Mare Tranquillitatis Lava Flows	3.57–3.88	
Archimedes		3.8 ?
Mare Serenitatis Lava Flows	3.65–3.85	
Mare Imbrium Impact Event (Includes Apennine Mts.)	3.85	
Mare Serenitatis Impact Event	3.87	
Mare Crisium Impact Event		
Mare Nectaris Impact Event	3.92	
Mare Fecunditatis Impact Event		
Mare Tranquillitatis Impact Event		
Oldest Mare Basalt obtained from a Highland Breccia	4.2	
Oldest Highland Rock found within a Breccia	4.356	

*Adapted from Grant H. Heiken et al., "Lunar Source Book," Cambridge University Press, 1991.

Table 17-II
Prominent Lunar Features

Apennine mountains	Lunar highlands near the South Pole
Archimedes crater	Mare Imbrium basin
Aristillus crater	Mare Imbrium lava flows
Clavius crater	Mount Piton
Copernicus crater	Plato crater
Eratosthenes crater	Tycho crater

DISCUSSION QUESTIONS

1. How do the geologic eras you devised differ from those used by lunar geologists? Comment on any differences.

2. Would the procedures used here work for developing either relative or absolute geologic chronologies for Mercury?

3. What other planets and/or satellites in the solar system would these procedures work for in determining geologic chronologies? What additional data would be needed? Explain your answer for each planet or satellite considered.

Figure 17-1. Partial Mosaic of the Moon.

(Taken by Scott Myers and courtesy Allan Morton, Central Arizona College, Signal Peak Campus)

Figure 17-2. Partial Mosaic of the Moon.

(Taken by Scott Myers and courtesy Allan Morton, Central Arizona College, Signal Peak Campus)

Figure 17-3. Partial Mosaic of the Moon.

(Taken by Scott Myers and courtesy Allan Morton, Central Arizona College, Signal Peak Campus)

Figure 17-4. Partial Mosaic of the Moon.
(Taken by Scott Myers and courtesy Allan Morton, Central Arizona College, Signal Peak Campus)

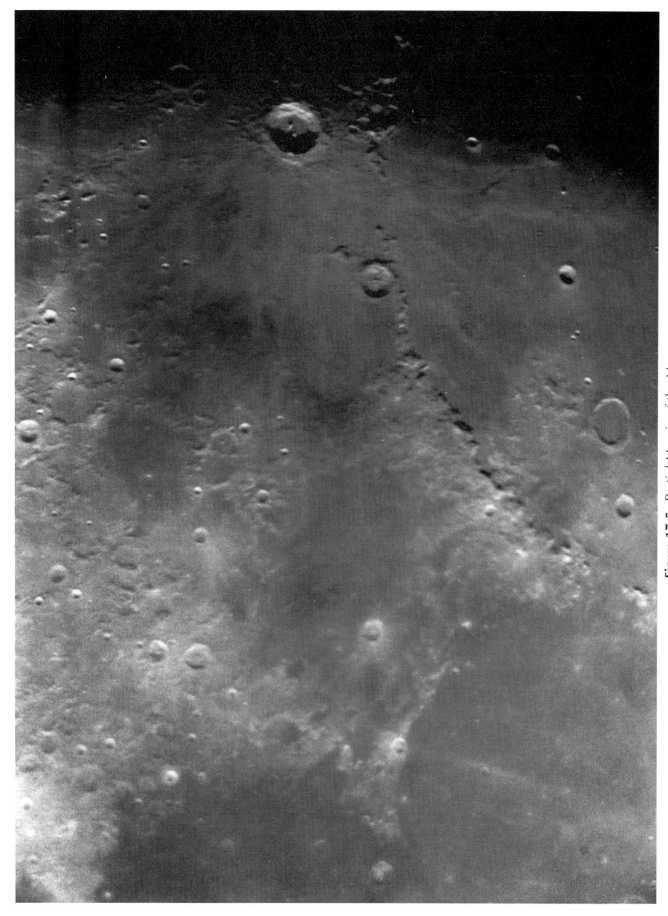

Figure 17-5. Partial Mosaic of the Moon.

(Taken by Scott Myers and courtesy Allan Morton, Central Arizona College, Signal Peak Campus)

Figure 17-6. Partial Mosaic of the Moon.

(Taken by Scott Myers and courtesy Allan Morton, Central Arizona College, Signal Peak Campus)

Figure 17-7. Partial Mosaic of the Moon.
(Taken by Scott Myers and courtesy Allan Morton, Central Arizona College, Signal Peak Campus)

Figure 17-8. Partial Mosaic of the Moon.

(Taken by Scott Myers and courtesy Allan Morton, Central Arizona College, Signal Peak Campus)

Figure 17-9. The Full Moon.
(Courtesy Allan Morton, Central Arizona College, Signal Peak Campus)

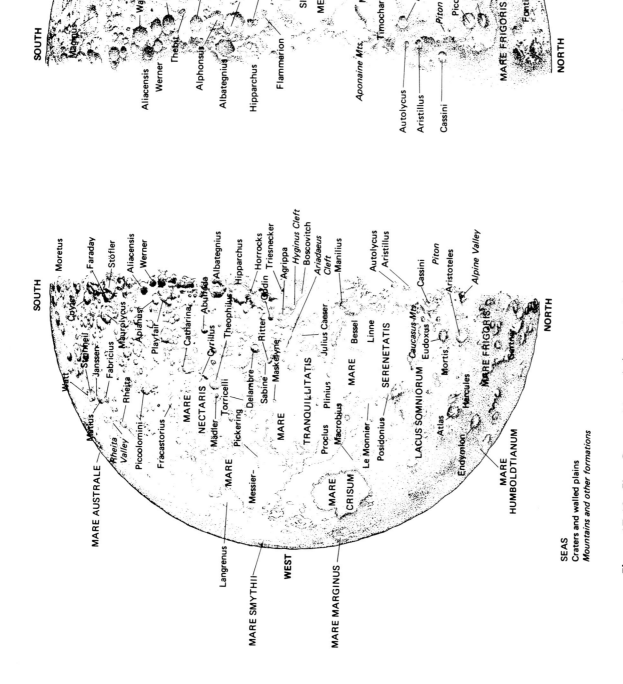

Figure 17-11. Third Quarter Moon (telescopic view).
(From AN INTRODUCTION TO ASTRONOMY, 2/e by Charles M. Huffer, Frederick E. Trinklein and Mark Bunge. Copyright © 1971, 1967 by Holt, Rinehart and Winston, Inc. Reprinted by permission of Holt, Rinehart and Winston.)

Figure 17-10. First Quarter Moon (telescopic view).
(From AN INTRODUCTION TO ASTRONOMY, 2/e by Charles M. Huffer, Frederick E. Trinklein and Mark Bunge. Copyright © 1971, 1967 by Holt, Rinehart and Winston, Inc. Reprinted by permission of Holt, Rinehart and Winston.)

18 Evidence of the Earth's Revolution

PURPOSE AND PROCESSES

The purpose of this exercise is to use the periodic Doppler shift of a stellar spectrum to determine the velocity of the earth as it revolves in its orbit about the sun, and to determine the value of the astronomical unit. The processes stressed in this exercise include:

Using Numbers
Identifying Variables
Controlling Variables
Formulation of Models
Questioning

INTRODUCTION

In 1842 Christian Doppler pointed out that when a light source approaches us or recedes from us, its apparent wavelength changes slightly. His explanation assumed that light behaved as if it were made up of waves and that the wavelength of light determines the color as we see it. (The longer wavelengths correspond to red light and the shorter to blue or violet light.)

If a light source approaches, the waves appear to be crowded closer together to give a shorter wavelength or a blue shift. Likewise, if the motion is away from the observer, the waves will be spread out to produce a red shift (Figure 18-1).

We can apply Doppler's principle to the light we receive from stars in order to determine the relative velocity between the star and the observer. Since the earth in its orbit is always approaching some stars and receding from others, lines in stellar spectra will show periodic Doppler shifts. We can use this fact to determine the velocity of the earth as it revolves about the sun. With knowledge of the earth's orbital velocity and its period, we can calculate the value of the astronomical unit.

The spectrogram in Figure 18-4 contains four separate spectra: At the top and bottom are iron spectra taken for comparison. (a) gives the spectrum of Arcturus on July 1, 1939; (b) is another spectrum of Arcturus, taken January 19, 1940. The comparison spectra of iron were taken by briefly illuminating an iron arc located on the telescope and are used as our rest-velocity standards. Iron is chosen because it has many easily identified lines of known wavelength. The two stellar spectra were taken six months apart at times when the earth most rapidly approached and receded from Arcturus. We assume that the velocity of Arcturus with respect to the sun is constant.

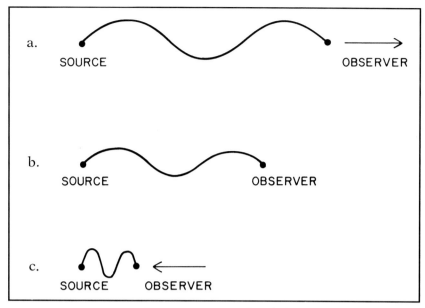

Figure 18-1. Doppler Shift.
(a) Red Shift: observer in motion away from the source.
(b) No Shift: observer at rest with respect to the source.
(c) Blue Shift: observer in motion toward the source.

PROCEDURE

1. A qualitative estimate of direction of motion is easily made for each spectrum by looking at several iron lines in the spectrum of Arcturus. Various iron lines have been identified for your convenience in Figure 18-3. To simplify your work, use only those lines in the spectra of Arcturus which are iron lines and identified in Figure 18-3. Note, however, that there are many other lines and several absorption bands as well.

   (a) What type of Doppler shifts do you observe in spectra (a) and (b)? Which is larger?

   (b) Since we are assuming that the star moves at a constant velocity, the apparent shifts must be because of the earth's motion. From this information you can tell whether Arcturus is approaching or receding from the sun. Which is it doing? Indicate your reasoning process by comparing relative velocities in a diagram such as 18-2.

2. (a) Measure the shifts of several prominent iron lines with respect to the corresponding iron comparison lines in each of the two spectra of Arcturus. Note that the narrow lines should be easier to measure accurately. A common method of measurement is to draw a straight sharp line through the two iron comparison spectra and across those of Arcturus as a reference. Try to es-

timate your measurements to a tenth of a millimeter. Using at least three lines, find the average shift for each spectrum.

   (b) We can measure the shift in wavelength of the lines directly in millimeters: however, we need a scaling factor, called *dispersion*, to convert these values to angstrom units. In general, dispersion tells us how "spread out" a spectrum is, and is expressed in angstroms per millimeter.

   Determine the dispersion of these spectra (which we assume to be linear and the same for all spectra on the plate) by measuring the distance in millimeters between two known lines of the iron comparison spectrum and dividing this value into the known wavelength difference of the lines. Do this calculation for several pairs of iron lines and find an average dispersion in angstrom/mm.

   (c) Convert your millimeter measurements to angstroms, noting whether the shift is toward the red or blue.

   (d) Calculate the value of relative velocity between the earth and the star for each spectrum using the Doppler equation:

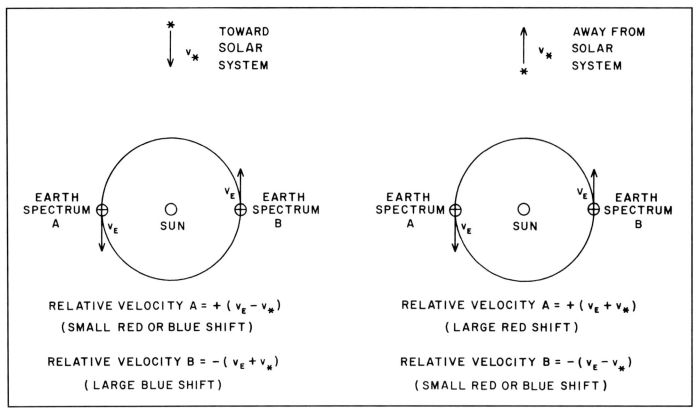

Figure 18-2. Relative Velocities of a Star Measured Six Months Apart.

$$v = c \, \frac{\Delta\lambda}{\lambda}$$

where v = relative velocity of source and observer
c = speed of light = 3.00×10^5 km/sec
$\Delta\lambda$ = apparent shift in wavelength
λ = wavelength of comparison line (at rest).

(e) Remembering that Arcturus has a constant velocity in the direction determined in part 1, calculate the orbital velocity of the earth by combining the equations for relative velocities given in Figure 18-2.

(f) Determine the velocity of Arcturus in km/sec with respect to the sun.

(g) We can find the radius of the orbit of the earth (the value of the astronomical unit) because we know the velocity of the earth and the period of its orbit (one year). The relation is expressed:

$$\text{velocity} = \frac{\text{distance}}{\text{time}}$$
$$= \frac{\text{circumference of orbit}}{\text{period}}.$$

By assuming a circular orbit, we have

$$v = \frac{2\,\pi\,R}{P}$$

where v = orbital velocity of the earth
R = radius of the orbit = 1 A.U.
P = period of the orbit.

Using your data, find the value of the astronomical unit assuming a circular orbit for the earth.

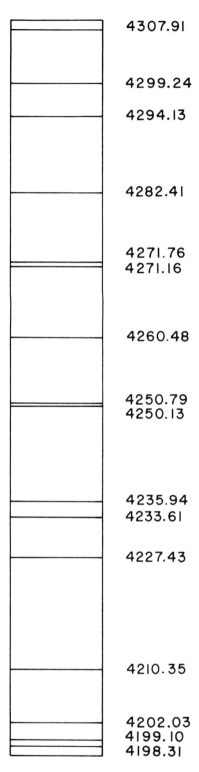

4307.91

4299.24

4294.13

4282.41

4271.76
4271.16

4260.48

4250.79
4250.13

4235.94
4233.61

4227.43

4210.35

4202.03
4199.10
4198.31

Figure 18-3. Line Identification for Iron Arc Comparison Spectra. (Not to the same scale as Figure 18-4.)

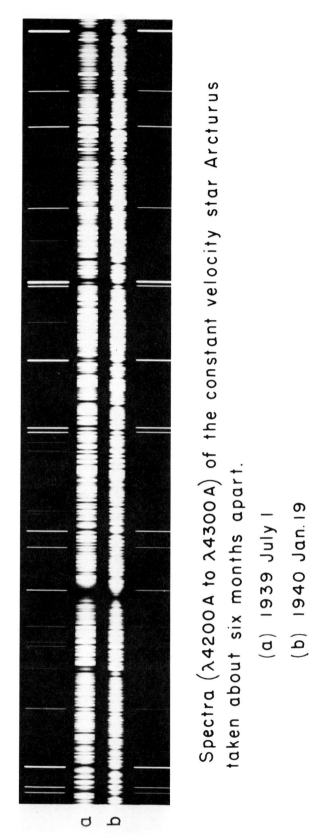

Spectra (λ4200 A to λ4300 A) of the constant velocity star Arcturus taken about six months apart.

(a) 1939 July 1

(b) 1940 Jan.19

a   b

Figure 18-4. (Courtesy of the Hale Observatories)

DISCUSSION QUESTIONS

1. In general, which stars does the earth DIRECTLY approach and recede from during the course of a year? Use a star chart to name one or two.

2. Assuming its present right ascension, what would the declination of Arcturus have to be in order for these spectra to reflect maximum velocity of revolution? Explain.

3. Can the earth's orbital velocity be determined accurately using spectra of stars from which it does not directly approach or recede? What is the necessary correction? Explain.

4. Arcturus, as seen with the naked eye, appears to be red in color. Why do we not see a change in color of the star as we change our direction from approach to recession with respect to the star?

19 Collecting Micrometeorites

PURPOSE AND PROCESSES

The purpose of this exercise is to collect micrometeorites and to use the collection to estimate the total infall of micrometeoric material over time.

The processes stressed in this exercise include:

Observing
Identifying Variables
Interpreting Variables
Using Numbers
Using Logic
Inferring

INTRODUCTION

Sky watchers report seeing about 5 or 6 "shooting stars" per hour when the skies are clear and dark. During periods of meteor showers, the hourly rate is even greater. Yet the number of large recoverable objects called meteorites is quite low. What happens to the materials that make up the bulk of observed meteors? As early as the 19th century some people speculated that the smaller objects that "burned up" in the atmosphere to produce the "shooting star" phenomena left behind small, microscopic materials that came to be known as micrometeorites. These small particles result from the ablation from larger meteors that plunge through the atmosphere and melt as a result of the heat generated when the meteors are slowed by atmospheric friction. These small pieces quickly cool into spherical shapes and, over a few days, slowly settle to the Earth.

This activity permits students to physically examine actual visitors from outside the Earth. While most students will never be able to actually collect macrometeorites, they will be able to collect micrometeorites, examine them under a microscope and, if they are persistent over time, to determine if there is any correlation between meteor showers and the amount of micrometeoric materials they collect.

Caution: Not all scientific investigators agree that the collection of micrometeorites in the manner described in this lab leads to reliable results.

PROCEDURE

1. COLLECTING MICROMETEORITES

There are a number of ways to collect micrometeorites. Choose one of the following methods which best suits your circumstances.

(a) Collecting with Magnets

Using small donut-shaped magnets, tie a piece of cord to the magnet and cover the magnet with small plastic freezer bag. Go out-of-doors and start collecting by allowing the magnet to trail close to the surface of sidewalk cracks, street gutters, near drain spouts, open parking lots and gravel driveways. (Or, following a rain storm, clean the material from the gutters of a house. Allow it to dry and drag a covered magnet through the material and transfer to a plate.) After you have gathered enough materials to be visible on the magnet bag, carefully pull the magnet out of the covering so that material collected falls into a paper plate or aluminum pan. Pour the material you collected through a screened sieve to remove large pieces and pour the smaller materials into a covered glass container.

(b) Collecting on Glass Surfaces

Place a thin layer of petroleum jelly on the surface of glass microscope slides or on the bottom of petri dishes. Place these objects in an outdoor area where they will not be disturbed for several days. In this manner you will gather non-magnetic and magnetic micrometeorites as well as spores and pollen grains.

(c) Collecting from Snow

Wait for 3 or 4 days after a snowfall. Using large containers take random samples of snow at different depths. Melt the snow and use a large funnel and coffee filters and filter the melted snow. Flatten the filter paper and allow it to dry.

2. EXAMINING THE COLLECTED MATERIAL

(a) Using a Microscope

Use a microscope and place a very small amount of the material collected onto a microscope slide. Bamboo skewer sticks can be used as tools to move the objects around under magnification. (Regular needles, pins and probes will not work because the iron meteoric materials will become magnetized if a magnet was used to collect them.) Examine the materials with a 25×–30× magnification. The small round spherules are micrometeorites! See Figure 19-1.

(b) Testing for Iron and Nickel

Nickel is common in iron-nickel meteorites but it is not common in fly ash. One diagnostic test that is relatively easy to perform is to determine if nickel is present in the magnetic spherules. Place a few of your spherules in a test tube and dissolve them with dilute nitric acid. Then add a few drops of ammonium hydroxide. A red-brown precipitate indicates the presence of iron. Add more ammonium hydroxide until the acidic solution has been neutralized. Pour the solution into a clean test tube and add a few drops of 1.0% solution of dimethyl glyoxime. A rose red precipitate indicates the presence of nickel.

3. EXTENTIONS

(a) Use 10 slides coated with petroleum jelly and allow them to sit out in several exposed locations for 24 hours. Collect the slides and count the number of spheres. Calculate how many microscope slides it would take to cover the surface of the Earth. The total

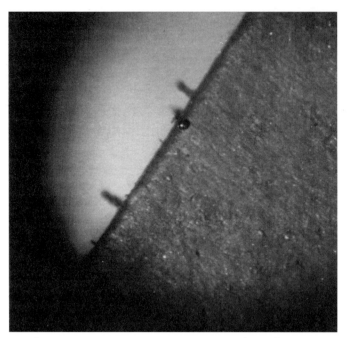

Figure 19-1. Micrometeorite as seen through a 25× magnified microscope. The darker background is the edge of a magnet. The small round sphere in the center of the field is a micrometeorite of approximately .15 mm (.006 inch) diameter. (Taken by Joe Toubes, 1973)

surface area of the Earth is 5.14×10^{18} cm^2. Measure the area of your microscope slide in square centimeters and divide this number into the total terrestrial surface area. Use the average number of meteoric spheres you found on your slides and multiply this times the number of slides it would take to cover the surface of the Earth. Some sources calculate that there may be as many as 200 million spherules landing daily on the Earth.

(b) If you have access to a precision electronic balance, collect a sample of micrometeorites from a measured amount of surface area and weigh it. Extrapolate this figure to calculate the amount of meteoric mass landing on the Earth each year.

(c) Start a long-term collecting project to see if the number of spherules collected is related to the occurrence of meteor showers. Most investigators believe that it takes several days following a shower for the particles to settle out of the atmosphere. The settling rate is likely accelerated by rain or snow falls as it is believed that the particles serve as nucleation particles for precipitation. See Figure 19-2 for a frequency distribution of micrometeoric frequency as a function of time of year.

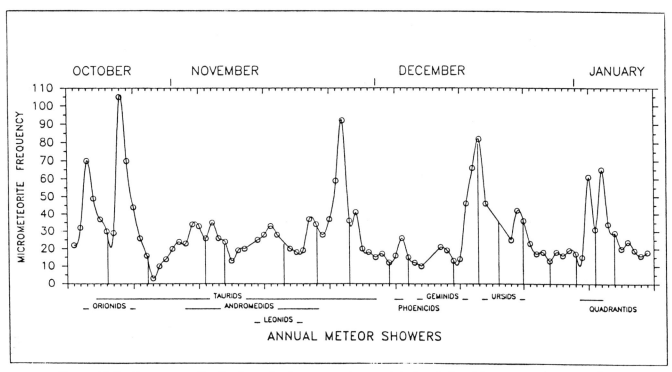

Figure 19-2. Frequency distribution of micrometeorites and their relationship to annual meteor showers. (R. Jones, "Chasing meteors with a microscope.")

DISCUSSION QUESTION

1. Why would the estimated mass of meteoric material collected by using a magnet likely produce a low estimate for the amount of annual infall?

20 Height of a Meteor

INTRODUCTION

The phenomena of "shooting stars" has appealed to humans from ancient times. From the time of Aristotle these events were associated with the atmosphere, but he failed to recognize their true nature. In the later 1790s two students at the University of Göttingen (after reading Chladni's definitive book on the subject) devised and conducted a crude, but successful experiment by observing meteors from two locations separated by several kilometers. From simple triangulation, they were able to calculate that meteors occurred about 80 kilometers above the earth.

Using two camera stations, separated on an east-west line and directed straight up, simplifies the geometry involved and is suggested here. With success, photographic records can be obtained and can later be processed to obtain the desired information.

PROCEDURE

Choose a date in the year during which a predicted meteor shower will occur. The list given on the preceding page is adapted from *Comets, Meteorites and Men.*[1]

Station two teams of photographers at least 5 kilometers apart. From careful ground measurements or from measurements made from topographic maps, determine the distance between the two stations. Position cameras so that they are photographing directly overhead. (See Exercise 6 "Astronomical Photography")

Arrange a set photographic schedule to insure that exposures from the two stations can be later matched. A suggested exposure of 14 minutes, followed by a one minute interval to rewind and reset the camera after each exposure is recommended.

Develop and inspect the negatives from the two stations. If you have been successful in "capturing" a meteor on equivalent exposures, print copies from each station to the same scale.

It will be necessary to identify two stars on the photograph to establish an image or plate scale. (See Exercise 12, "Image Size-Focal Length Relationship.") Choosing

1. Adapted from Peter Lancaster Brown, *Comets, Meteorites and Men* (New York Taplinger), 1974, pp. 244–245.

stars directly north and south of each other and photographing directly overhead in the mid-northern latitudes minimizes the problems of geometry.

Using your photographs measure the linear separation between the streaks on the two photographs. With your established plate scale, calculate the angular separation between the two streaks. From geometry calculate the height of the meteor. (See Figure 20-1.)

Optional Activity

If your photographs are not successful, use those included here. These photographs were taken on December 13, 1975 during the Geminid shower between 11:14–11:30 pm CST. The bright star in the center of the photograph is Capella. Down and to the right is Epsilon Aurigae, and the next double streaks are Eta and Zeta Aurigae. The two observers were located 3.2 kilometers apart.

Using the information in Table 20-II, establish an image or plate scale and determine the height of the meteor.

Table 20-I
The Major Annual Meteor Showers

	Date of Peak Activity	Radiant Coordinates		Duration of Dectectable Meteors	Duration of Peak Days	Expected[1] Hourly Rate
		R.A.	Dec.			
Quadrantids	Jan. 3	15h 24m	+50°	1–4 Jan.	0–5	50
Corona Australids	Mar. 16	16h 20m	–48	14–18 Mar.	5	5
Virginids	Mar. 20	12h 40m	00	Mar. 5–Apr. 2	20	5
Lyrids	Apr. 21	18 h08m	+32	19–24 Apr.	2	10
Eta Aquarids	May 4	22h 24m	00	Apr. 21–May 12	10	20
Ophuichids	June 20	17h 20m	–20	17–26 June	10	20
Capricornids	July 25	21h 00m	–15	July 10–Aug. 20	20	20
Southern Delta Aquarids	July 29	22h 36m	–17	July 21–Aug. 15	15	20
Northern Delta Aquarids	July 29	22h 36m	00	July 15–Aug. 18	20	10
Pisces Australids	July 30	22h 40m	–30	July 15–Aug. 20	20	20
Perseids	Aug. 12	03h 40m	+58	July 25–Aug. 17	5	50
Kappa Cygnids	Aug. 20	19h 40m	+55	18–22 Aug. 5	3	5
Orionids	Oct. 21	06h 20m	+15	18–26 Oct.	5	20
Southern Taurids	Nov. 1	03h 28m	+14	Sept. 15–Dec. 15	45	5
Northern Taurids	Nov. 1	03h 36m	+21	Oct. 15–Dec. 1	30	5
Leonids	Nov. 17	10h 08m	+22	14–20 Nov.	4	varies
Phoenicids	Dec. 5	01h 00m	–55	Dec. 5	0.5	50
Geminids	Dec. 13	07h 32m	+32	7–15 Dec.	6	50
Ursids	Dec. 22	14h 28m	+80	17–24 Dec.	2	5

1. These numbers are averages of hourly rates based on numbers that increase and decrease from one year to the next. Annuals, calendars and the monthly almanac or calendar sections of *Sky and Telescope* and *Astronomy* will give current estimates of the hourly rate for a given shower or date.

Table 20-II
Position of Stars in Figure 20-1

Star	RA	DEC
Capella (α Aur)	5h 17m	46°00′
Epsilon Aur	5h 02m	43°50′
Eta Aur	5h 06m	41°14′
Zeta Aur	5h 02m	41°05′

Figure 20-1. Photographs of a Geminid Meteor Taken from Two Locations Separated by 3.2
Kilometers.
(Courtesy: Darrel Hoff and Thomas Wagner, University of Northern Iowa)

DISCUSSION QUESTIONS

1. What would be the advantages and disadvantages of having the two viewing locations closer together? Farther apart?

2. Suggest a process by which one could time the duration of a meteor.

3. How might one estimate the magnitude of a meteor?

21

Mercury's Rotational Period

PURPOSE AND PROCESSES

The purpose of this exercise is to determine the rotational period of the planet Mercury using reflected radar data. The processes stressed in this exercise are:

Using Numbers
Interpreting Data

INTRODUCTION

Determining the basic properties of the planets (sizes, masses, motions) is an important first step for the observational astronomer. This information can be used to build a more complete understanding of the solar system. Sometimes direct observation suffices, but in other cases it provides misleading information. The rotational period of Mars was successfully determined in this manner. However, similar attempts for Mercury made by Bessel, Schiaparelli, and others produced conflicting results. Bessel deduced a period of about 24 hours, but long term observations by Schiaparelli (and apparently confirmed by others) placed its period at 88 days. This is the same as Mercury's period of revolution, and was often described as being in "captive rotation" with one face always towards the sun.

The first attempts were made to use spectroscopy about 1900 to determine rotation periods of planets. It was first applied by Keeler to the rings of Saturn, and attempts were made by Slipher and others to determine Mercury's rotational period. The slit of a spectrograph was laid parallel to the equator of the planet. The lines from the receding edge were red-shifted while those from the approaching edge were blue-shifted, obeying the classic Doppler formula. The results obtained by this method indicated that Mercury took several days to rotate, but precise measurements could not be made. (See Exercise 18. "Evidence of the Earth's Revolution" and Exercise 24. "Solar Rotation" for other applications of this method.)

A much more powerful method became possible during the 1960s when radar signals were successfully bounced from planetary surfaces. Pettengill, Dyce, and Shapiro produced an accurate rotational period for the planet Mercury using this method. In 1965 they used the 1000-ft. radio telescope at Arecibo, Puerto Rico to beam a series of 0.0005 second and 0.0001 second radar pulses at 430 megahertz toward the planet. Since the round trip travel time of the pulses was much greater than the pulse length they could see how the pulses were broadened by reflection from the rotating planet. Frequency shifts also resulted from the relative motions of the planet and the earth's rotation, but these were corrected by using careful timing and computer compensation.

Figure 21-1 shows that when a radar signal is reflected from a rotating spherical planet the echo is spread out in time as well as in frequency. The echo first returned is from the sub-radar point. After a small time delay the echo is received from a ring-shaped area centered on this point. That part of the signal returned from the approaching edge will be returned with an increase in frequency ("blue-shifted") and that point returned from the receding edge will be returned with a decrease in frequency ("red-shifted").

126

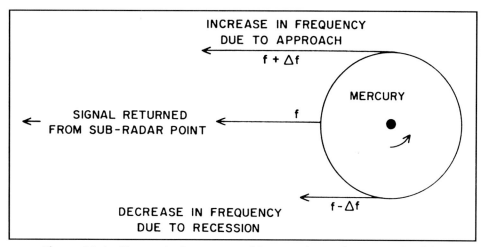

Figure 21-1. Change in Frequency of Returned Signal as a Result of the
Rotation of Mercury.

Figure 21-2 shows the spectrum of the radar echo for five different time delays (Δt). Note that the longer
the time delay, the broader is the return signal in frequency. This is because successive signals return from
farther and farther from the sub-radar point. The portion of the planet rotating toward the earth causes
the return signal to have an increase in frequency (+) and the portion rotating away has a decrease (−).
This increase or decrease obeys the Doppler law.

In principle it ought to be easy to determine the rotational velocity of Mercury's limb, and (by knowing
the planet's circumference) to calculate the rotational period. However, the echo weakens towards the
edge of the planet and the signal return from the limb is unobtainable. We will use the echo from a ring
intermediate between the sub-radar point and the limb to calculate a line-of-sight component of
Mercury's rotational velocity and from this find its true rotational velocity.

Examine Figure 21-3. Note that in Figure 21-2 each signal is labeled with its time delay in microseconds
(μs) (1 microsecond = 10^{-6} seconds). It is easy to calculate the distance d that any delayed beam has traveled beyond that sub-radar point by multiplying half the delay time by the speed of the radar wave (the
speed of light). This information is needed to obtain the line-of-sight velocity (V_o) in order to get the true
rotational velocity (V).

PROCEDURE

1. Choose one of the time delayed signals in Figure 21-2
   and calculate

   $$d = \frac{1}{2} c \, \Delta t \qquad \text{Equation 1}$$

   where d is the distance in meters
     Δ is the time delay in μs
     c = 3×10^8 m/sec.

2. In Figure 21-3, the lengths x and y are given by

   $x = R - d$ Equation 2
   $y = (R^2 - x^2)^{1/2}$ Equation 3

   where R is the radius of Mercury = 2.420×10^6 m.
   Calculate x and y.

3. Using the previously selected signal from Figure 21-2
   find V_o, the observed line-of-sight component of the
   rotational velocity at some point indicated in Figure
   21-3. The Doppler equation is generally stated in
   terms of a change in wavelength (Δλ) relative to the
   "rest" wavelength (λ), but it can also be stated in
   terms of frequency (f)

   $$\frac{\Delta f}{f} = \frac{V_o}{c} \qquad \text{Equation 4}$$

   where Δf is the change in frequency

     f is the frequency of the transmitted signal
       = (430 megahertz = 4.3×10^8 hertz)
   V_o is the observed velocity

   c is the speed of the radar wave.

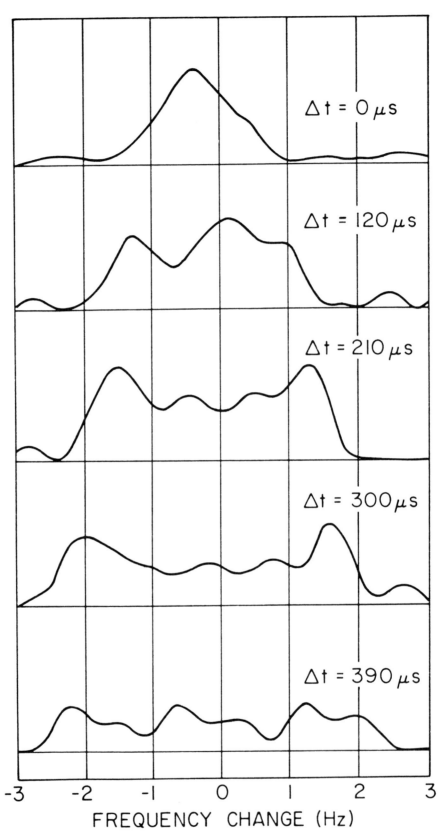

Figure 21-2. Spectrum of Radar Pulses Returning from Mercury Made on
August 17, 1965.
(Adapted from R. B. Dyce, G. H. Pettengill, and I. I. Shapiro,
Astronomical Journal, 72, 351 (1967); by permission of the American
Astronomical Society. Copyright ©1967 by the American Astronomical
Society.)

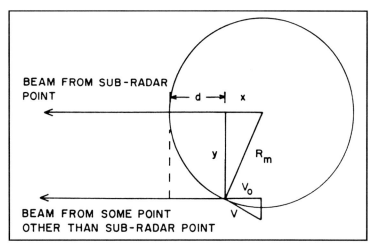

Figure 21-3. Geometry for Determining Mercury's
Rotational Period from Reflected Radar Pulses.

Examine the selected radar signals in Figure 21-2 and mark with a pencil the points to the left and right of center where the relative power *begins* to drop down to the baseline. Read off the frequency change at each of these points as accurately as you can. Disregarding algebraic signs, average the results from the two shoulders. The actual Doppler frequency shift, Δf, is half the value as this is a reflected signal and the radar pulse is shifted "going in" and again in "coming out." Calculate V_0 in meters per second.

4. From the line-of-sight component V_0, calculate V, the foreshortened rotational velocity. Inspect Figure 21-3 and note that the triangle containing x, y, and R is similar to the triangle containing V_0 and V. Hence,

$$\frac{V}{V_0} = \frac{R}{y} . \qquad \text{Equation 5}$$

Calculate V from equation 5. The result is the true rotational velocity in meters per second.

5. Calculate Mercury's rotational period by dividing V into the circumference of Mercury $= 1.520 \times 10^7$ m.

DISCUSSION QUESTIONS

1. What are some possible reasons why Schiaparelli observed an 88-day rotation period?

2. In order to minimize the motions of the two planets, what would be the ideal time to make this set of observations? Explain. (If old copies of the *American Ephemeris and Nautical Almanac* are available check to see if your predicted time was that used.)

3. Venus' rotation is considerably slower than Mercury's. If this method was used on Venus, would the frequency change be larger or smaller than that for Mercury? (Keep in mind that the planets are not the same size.) Explain.

22 Measuring the Diameters of Pluto and Charon

PURPOSE AND PROCESSES

The purpose of this exercise is to use occultation data from the Pluto-Charon system to determine the diameter of each of the two bodies. The processes stressed in this exercise include:

Using Numbers
Controlling Variables
Interpreting Data
Formulation of Models
Using Logic
Inferring

INTRODUCTION

Pluto is the only planet discovered in the 20th century. In 1930, Clyde Tombaugh culminated a year-long search of photographic plates at Lowell Observatory. He discovered a small image which showed motion against the background of stars, matching the predicted motion of a new outer planet. The new planet was named Pluto after the ancient Greek god of the underworld. Further study revealed that Pluto's orbit was quite eccentric and had an average distance of about 40 astronomical units from the Sun. At that distance, the planet requires 248 years to complete one orbit about the Sun.

After the planet was discovered, attempts were made to determine its physical properties, including its size and mass. Obtaining fundamental information about the planet proved to be a difficult task. Pluto is a dim object of only about 15th magnitude. Even in the largest telescopes, Pluto does not appear as a disk. As a result, directly measuring its angular size was not possible. Through the years, many attempts to determine its size and mass led to a variety of estimates. In 1950, Gerard Kuiper used a planetary occultation of a star to estimate Pluto's size as approximately 6000 km. This value appeared in the literature for a quarter century, but most astronomers acknowledged that it was only an informed guess.

Accurately determining the mass of a planet requires that the planet interact with the gravity of another object such as a nearby planet or a moon. Accurate masses for most planets are determined by their gravitational interaction with one of their moons. Pluto's mass therefore was uncertain because it did not have a detectable moon. All of this changed in 1978. While studying U.S. Naval Observatory photographic plates of Pluto, James Christy discovered a bump on the side of Pluto's images. Closer examination of the images revealed that the bump was the image of a satellite, located so close to the planet that the image of the moon and the planet blurred together. Christy named the moon Charon after the mythological boatman who carried the dead souls across the River Styx to the underworld dominated by Pluto. It was soon determined that the moon orbited the planet with a period of 6.39 days—exactly the same as the rotation period of the planet! From a knowledge of this period and the size of Charon's orbit, the mass of the planet-moon system was soon determined. (See Exercise 14, "Determining the Mass of the Moon" for an activity on determining the mass of an object from the motion of a satellite around the object.)

The problem of determining the sizes of the moon and the planet was still difficult. However, astronomers soon made a discovery about how Charon orbited Pluto that assisted in solving the size problem. Charon revolves around Pluto at almost a right angle to the path of the planet around the sun (see Figure 22-1). When Earth is in the same plane of Pluto-Charon's orbit, Charon alternately passes behind

130

or in front of Pluto. It was soon recognized that these events could be used to accurately determine the diameters of the two bodies. This parallel geometry takes place only twice in Pluto's 248 year orbit about the Sun. By good fortune, the required geometry to observe the movement of the moon in front or in back of the planet occurred between the years 1985 and 1990.

An occultation occurs when a planet blocks the light from a more distant object. When a satellite passes in front of a planet, the event is called a transit. We can use information about an occultation of Charon by Pluto to determine the respective sizes of Pluto and Charon. Figure 22-2 shows such an occultation about ready to start. First contact occurs when the limb of Charon just touches the limb of Pluto. Second contact occurs at the moment the moon completely disappears behind Pluto. The length of time between these two events is a function of the velocity of the satellite and its diameter. We can reason similarly that the time interval between first contact and when the moon begins to emerge from behind the planet (third contact) is a function of the moon's velocity and the diameter of Pluto. We can measure these time intervals and use the known velocity of Charon to measure the diameter of both Pluto and Charon.

It is impossible to actually see Charon and Pluto separately except in the largest telescopes. Direct observation therefore cannot be used to time the instant when Charon begins to slip behind the planet or when it reappears. Astronomers instead observe the amount of light being received from the system. When the two objects are separated from each other, the amount of light received from the system is greater than when one of the objects is blocking the other. Sensitive instruments called photometers are attached to a telescope and the amount of light falling on the photometer is carefully measured. The time also is recorded for each observation.

The amount of light entering the photometer decreases when an occultation or transit starts. Astronomers produce what is known as a light curve by plotting the decrease in the amount of light (here called a delta magnitude) as a function of time. Figure 22-3 shows an idealized light curve from an occultation event. We can determine the length of an occultation from such a light curve. We can then calculate the diameters of the planet and moon from the elapsed times and the known velocity of the moon in its orbit. This step requires using the familiar formula that relates the distance (d) an object travels to its velocity (v) and time (t) of travel: $d = vt$.

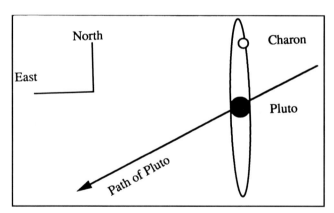

Figure 22-1. Every 124 Years, the Plane of Charon's Orbit Lines up with the Earth and the Satellite Alternately Crosses in Front of or Behind Pluto.

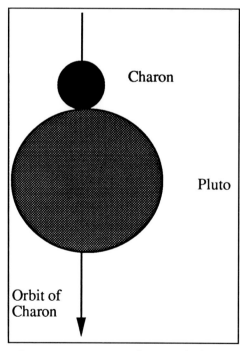

Figure 22-2. An Occultation of Charon About to Start.

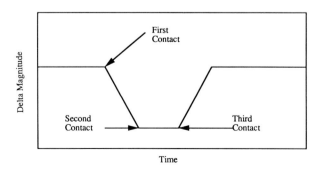

Figure 22-3. Sample Light Curve of Pluto Occulting Charon.

---| PROCEDURE |---

1. Table 22-I shows a reconstructed set of data of Pluto occulting Charon. The first column gives the time of the observation. The second column lists the change in the amount of light, expressed as *delta magnitude*. Minus values indicate a decrease in the amount of light being measured by the photometer.

2. Use the graph paper provided and plot each delta magnitude as a function of time. After completing the plot, use a straight edge to construct the light curve.

3. Using the data sheet provided, record your time estimate of the first, second and third contacts. From your graph, determine the elapsed time between the first contact and the second contact. This is the length of time required for Charon to disappear behind the planet. Record this value.

4. From your graph, determine the time from the first contact until the third contact. This is the length of time required for Charon to traverse the diameter of Pluto itself. Record this value on your data sheet.

5. According to recent estimates, Charon has an orbital radius (r) of 19,130 km. Assuming that the orbit of Charon is a circle, calculate the circumference (C) of Charon's orbit using the formula $C = 2\pi r$. Record this value in your data sheet.

6. It takes 6.387 days or 153.3 hours for Charon to complete one orbit around Pluto. Assume again that the orbit of Charon is a circle and calculate its orbital velocity. Record this value in your data sheet.

7. Multiply the orbital velocity of Charon by the elapsed time recorded in step number 3. This will be your estimate for the diameter of Charon. Record the value in your answer sheet.

8. Multiply the orbital velocity of Charon by the elapsed time recorded in step 4. This is your estimate for the diameter of Pluto. Record this value.

Table 22-I

Photometric Observations of Pluto Occulting Charon. Reconstructed data from February 18, 1987 mutual event

Time (UT)	Delta Magnitude
13:30	−0.010
13:45	−0.001
14:00	+0.012
14:15	−0.011
14:30	−0.033
14:45	−0.050
15:00	−0.097
15:15	−0.128
15:30	−0.155
15:45	−0.180
16:00	−0.218
16:15	−0.220
16:30	−0.221
16:45	−0.219
17:00	−0.220
17:15	−0.199
17:30	−0.150
17:45	−0.130
18:00	−0.092
18:15	−0.049
18:30	−0.024
18:45	−0.005
19:00	−0.006
19:15	−0.002
19:30	−0.001

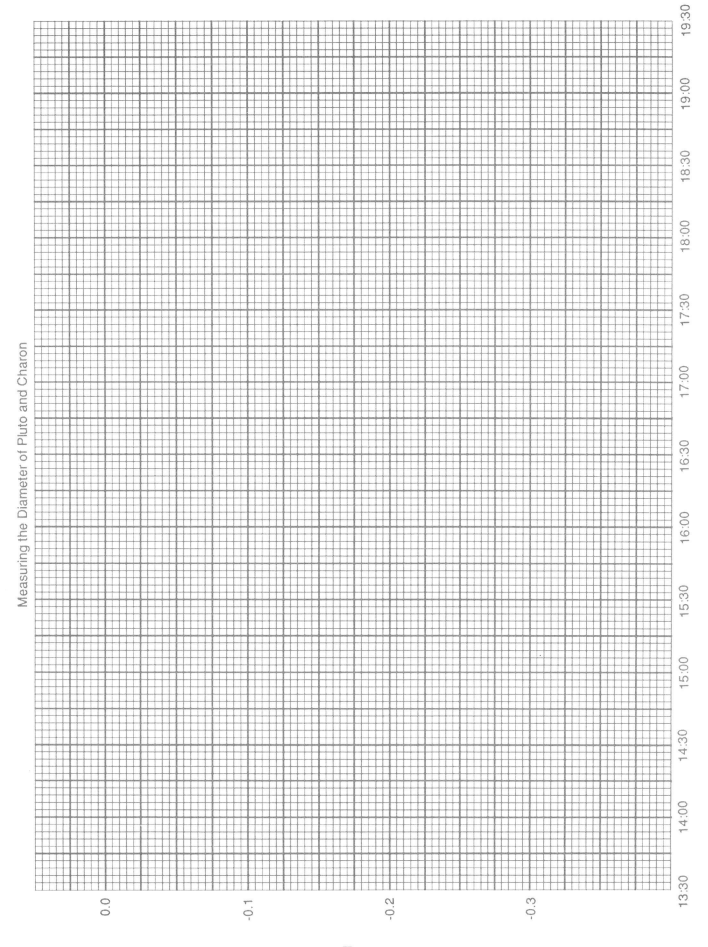

Measuring the Diameter of Pluto and Charon

Delta Magnitude

0.0   -0.1   -0.2   -0.3

13:30 14:00 14:30 15:00 15:30 16:00 16:30 17:00 17:30 18:00 18:30 19:00 19:30

22. Measuring the Diameter of Pluto and Charon

1. The time of contacts occurred at:

   First _____ hours   Second _____ hours   Third _____

2. The elapsed time between first and second contact was _____ hours. (Express answer in hours and decimal fractions.)

3. The elapsed time between first and third contact was _____ hours. (Express answer in hours and decimal fractions.)

4. The orbital circumference of Charon is _____ kilometers.

5. Assuming that the orbit of Charon is a circle, its orbital velocity is _____ km/hr.

6. My calculated diameter of Charon is _____ km.

7. My calculated diameter of Pluto is _____ km.

DISCUSSION QUESTIONS

1. What are some possible sources of error in this experiment?

2. When Charon transits the planet, there is a greater decrease in the amount of light measured from the system than when Charon passes in back of the planet. What might be some reasons for this difference?

3. How might measurements made from above the Earth's atmosphere by the Hubble Space telescope help in determining the fundamental properties of the Pluto-Charon system?

4. If Pluto had a gaseous atmosphere, how might this atmosphere change the shape of an occultation light curve and how might this change the calculations resulting from it?

23 | Determining the Velocity of a Comet

INTRODUCTION

Comets have fascinated humankind for centuries. In ancient times these objects were regarded as omens of doom and the appearance of bright comets often shaped human events. For example, the appearance of a very bright comet in 1066 may well have influenced the outcome of the Battle of Hastings. The losing Saxons, under the leadership of Prince Harold, regarded the comet's appearance as a bad omen and may have been adversely affected. The bright comet of 1456 appeared at the same time as a Turkish invasion in western Europe and the two events were believed to be related. Pope Calixtus II ordered public prayers for deliverance from the evil effects of both the comet and the invaders.

Work done by Edmond Halley, in the early 18th century, showed that comets travel in orbits controlled by the gravity of the Sun. Halley was able to show that some comets seen in the past were really the same comet returning at predictable intervals. In fact, he showed that the comets of 1066 and 1456 were the same comet. Examination of the movement of the comet convinced him that this comet should return to the vicinity of the Sun at 75 year intervals. His successful prediction of its return in 1758 gave a powerful boost to the acceptance of Newtonian physics. The comet bears Halley's name today.

Comets are no longer regarded as bad omens. A large number of them have been discovered and their orbits charted in the last few centuries. Approximately 6–12 new comets are discovered each year and others are observed on their periodic return to the vicinity of Earth. Comets are usually named for their discoverers. If more than one individual makes an independent discovery, the comet is named for its first (up to three) co-discoverers. The comet you will investigate in this activity, Comet Kobayashi-Berger-Milon, was discovered in 1975.

Figures 23-1 and 23-2 are pictures of Comet Kobayashi-Berger-Milon. The original photographs were taken at the University of Northern Iowa. The exposures were ten-minutes long and were made with an SLR camera piggy-backed on a Celestron–8 telescope. A 50–mm lens was set at f/1.8 and Tri-X film was used. In one exposure, the stars were tracked and the comet was allowed to trail. In the second photograph, the comet itself was tracked. Can you tell which picture is which?

Figure 23-1. Comet Kobayashi-Berger-Milon (1975h) Photographed at 11:30–40 P.M. (CDT) on the Night of July 24/25, 1975.

Figure 23-2. Comet Kobayashi-Berger-Milon (1975h) Photographed at 1:30–40 A.M. (CDT) on the Night of July 24/25, 1975.
(Courtesy Darrel Hoff and Tom Wagner, University of Northern Iowa)

You will make measurements from these photographs to determine the position of the comet at the time of each exposure. From these positions it will be possible to calculate the angular distance that the comet moved in the two-hour interval between the pictures. Using this angular distance and the distance to the comet it will be possible to calculate the linear distance that the comet has moved in two hours. Once this linear distance is known, you will calculate the comet's velocity. In order to complete this activity, it might be useful for you to review the concept of plate scale as found in Exercise 16, "The Moon's Sidereal Period". The small angle formula should be reviewed as well. The latter formula relates the angular movement of an object in radians (θ) to its linear movement (s) and its distance (r):

$$s = r\,\theta$$

PROCEDURE

1. Figure 23-1 was taken at 11:30–40 P.M. (CDT) on the night of July 24/25, 1975 and Figure 23-2 was taken two hours later. The fuzzy spot in the right central section of each picture is the comet. The brightest star in the field is Omicron Draconis, found to the left of the comet. Directly north of Omicron Draconis is SAO 029766. Figure 23-3 shows a section of a Smithsonian Astrophysical Observatory chart covering the same area of the sky. You will use the known coordinates of these two reference stars to establish a plate scale. You will then use your plate scale to determine the angular movement of the comet in the time interval between the two exposures.

2. Examine Figures 23-1, 23-2 and 23-3 and identify the comet and the two reference stars on each of the figures. Keep in mind that the scale of the drawing is not the same as the scale of the photographs. Differences in scale are a common problem in astronomical work and often require a very careful examination of visual records to insure that you are working with the correct celestial objects.

3. The coordinates of the two reference stars are as follows:

Star	Right Ascension	Declination
Omicron Draconis	16^h 0^m 57^s	58° 41′ 54″
SAO 029766	16^h 0^m 58^s	59° 46′ 04″

   a. Find the difference in declination between the two stars in seconds of arc. Record this value on your data sheet. Remember that there are 60 minutes of arc in a degree and 60 seconds of arc per minute of arc.

   b. Use a millimeter ruler and measure the linear distance between the images of Omicron Draconis and SAO 029766 on Figure 23-1 or Figure 23-2. Record this value to the tenth-millimeter.

   c. The information obtained from the last two steps will be used to establish a plate scale. Divide the

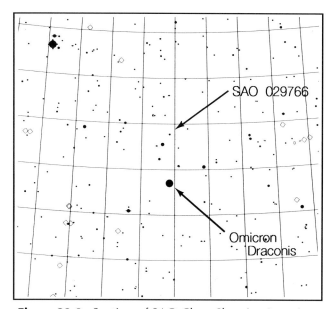

Figure 23-3. Section of SAO Chart Showing Locations of Omicron Draconis and SAO 029766.

angular separation between the two reference stars in seconds of arc (step 3a) by the linear photographic distance between the two reference stars (step 3b). The result of this division is your plate scale for these photographs expressed as the number of ″/mm of arc. Record this plate scale on your data sheet.

   d. Use a piece of tracing graph paper and mark the position of four or five of the brightest stars found on Figure 23-1. Carefully mark the position of the comet on this tracing.

   e. Transfer this tracing to Figure 23-2 and align it with the same reference stars. When your reference stars are aligned with the star images on the second picture, carefully mark the second position of the comet image on your tracing paper

4. a. From your tracing, use a millimeter ruler to measure the distance the comet has moved in the two hours that elapsed between the two pictures. Record this measurement in your data table.

    b. Use the plate scale from step 3c and convert the linear measurement from step 4a into angular measurement. How many seconds of arc did the comet move across the sky in two hours? How many seconds of arc did it move in one hour? Record this hourly angular velocity in the data table.

5. On the night these photographs were taken, the comet was 0.609 astronomical units (au) from the Earth, according to information supplied by the Central Bureau for Astronomical Telegrams at the Smithsonian Astrophysical Observatory. Convert this distance into kilometers. (Remember that there are 149,000,000 km/au.) Record this value in your data table.

6. During these exposures the comet was essentially moving from right to left across the sky and therefore its observed angular movement was a direct reflection of its linear movement. Use the hourly angular velocity of the comet from step 4b, the distance to the comet from step 5 and the small angle formula, calculate how far the comet moved in one hour. Record this value in your data table. Convert this velocity into km/sec by dividing by 3600.

23. Determining the Velocity of a Comet

1. The difference in declination of Omicron Draconis and SAO 029766 is _____ ″.

2. The linear distance between Omicron Draconis and SAO 029766, as measured from Figure 23-1,

   is _____ mm.

3. The plate scale of the photographs (linear distance/angular distance) is _____ ″/mm.

4. The distance between the comet images on the two photographs is _____ mm.

5. Using the plate scale (answer 3 above), and the distance between the two comet images (answer 4 above), the

   comet moved _____ ″ in two hours. Dividing this answer by two computes the comet's angular

   velocity as _____ ″/hr.

6. The distance to the comet is 0.909 au. This distance converted into kilometers is _____ km.

7. Using the distance to the comet in kilometers and its calculated angular velocity, the calculated velocity of the

   comet is _____ km/hr. Dividing this answer by the number of seconds in an hour (3600)

   computes the comet's velocity as _____ km/sec.

Use this space and the reverse side to show all calculations and to answer assigned discussion questions. Attach additional sheets if necessary.

DISCUSSION QUESTIONS

1. What are some possible sources of error in this experiment?

2. At the time of the photographs, the comet was 0.53 au from the Sun. At this distance from the Sun, the escape velocity from the solar system is 58 km/sec. How does this value compare with your measured velocity? What is the significance of this comparison?

PART III

The Sun, Stars and Galaxies

24

Solar Rotation

PURPOSE AND PROCESSES

The purpose of this exercise is to examine evidence that the sun rotates and to determine its approximate period of rotation. The processes stressed in this exercise include:

Using Numbers
Observing
Identifying Variables
Controlling Variables
Interpreting Data
Inferring
Predicting

INTRODUCTION

In the early 1600s Galileo first observed the sun with a telescope and discovered sunspots. This discovery was very disturbing to his contemporaries because the sun was thought to be a perfect "unblemished" celestial body. In his classic *Letters on Sunspots* he showed by rigorous argument that they were located on the surface of the sun and that their motion was evidence of solar rotation, and he produced an estimate of its period of rotation. In one of the earliest quantitative determinations about the sun, Galileo wrote, ". . . they [sunspots] have in common a general uniform motion across the face of the sun in parallel lines. From special characteristics of this motion, one may learn that the sun is absolutely spherical, that it rotates from west to east around its center, carries the spots along with it in parallel circles, and completes an entire revolution (sic) in about one lunar month."[1] It was not until the late 1800s, however, that an English amateur astronomer, R. C. Carrington, reported that sunspots at different solar latitudes require different times to complete one rotation.

We know now that sunspots are large areas which are relatively cooler and therefore darker than their surroundings. They are seen to last for a few days to several weeks and often are accompanied by large outbursts of optical, radio, x-ray and charged particle radiation.

Another way to determine the solar rotation rate is to use a spectrum taken with the spectrograph slit aligned along the solar equator. Since one limb is approaching us, one end of the solar lines will show a blue Doppler shift; likewise the end of the lines from the other limb will be red shifted. The part of the line originating from the center of the disc will not be shifted because the rotational velocity is transverse or perpendicular to our line of sight (Figure 24-1). As a result the lines are slightly tilted with respect to the lines originating in our own atmosphere. The amount of tilt is proportional to the difference in the velocities of the two limbs, and the period can be calculated. The periods of rotation at different solar latitudes can be obtained from spectra taken at those latitudes.

---

1. Drake, Stillman (translator). *Discoveries and Opinions of Galileo* (New York: Doubleday, 1957), p. 106.

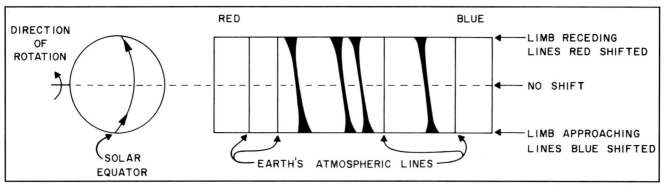

Figure 24-1. Orientation of the Sun and Solar Spectral Lines.

PROCEDURE

This exercise consists of three parts: (1) Examining a series of sunspot photographs to determine the direction and approximate period of solar rotation; (2) Investigating solar rotation based on the examination of spectra of the approaching and receding limbs of the sun at the equator; and (3) Comparing spectra taken at several solar latitudes. Not all parts need be done; check with your instructor to determine which parts to do.

You might also want to look at Exercise 4 "Observing Exercise III: The Sun" for suggestions on observing the sun and obtaining your own sunspot drawings or photographs.

1. SUNSPOT PHOTOGRAPHS

Figure 24-4 shows a series of photographs taken at the Hale Observatories during March and April of 1947.

(a) Cover a photograph with a sheet of tracing paper and carefully trace the edge of the sun. Sketch the outlines of the prominent spots, and record the date of the picture.

(b) Move your sketch to successive photographs until you find one that your spot tracing lines up with. Record the date and find the period.

(c) Repeat the above procedure with at least two more sets of photographs and find an average period of rotation.

(d) The photographs have the north pole of the sun at the top. Which direction is the sun rotating?

This method leads to determinating the sun's *synodic period,* an apparent period of rotation resulting from a combination of the sun's rotation and the earth's orbital motion.

2. SOLAR SPECTRUM

Figure 24-6 is a photograph of a small part of the solar spectrum taken along the solar equator. You can see two types of lines: narrow lines formed by oxygen in the earth's atmosphere; and the wide "fuzzy" solar lines. Close examination will show that the solar iron lines are slightly tilted with respect to the oxygen lines, which are not Doppler shifted at all. (You might want to review the Doppler effect and some of the details of measuring spectra in Exercise 18, "Evidence of the Earth's Revolution.")

(a) To make measurements easier, carefully draw a sharp line along and through the center of each solar line. The accuracy of your results will depend in part on being sure this line is at the center of the solar line along its entire length.

(b) The atmospheric oxygen lines are vertical and they provide excellent references for measuring the tilt (Δx) of the solar lines: simply measure from the top of an oxygen line to the top of a solar line, and from the bottom of the same oxygen line to the bottom of the solar line. The difference is the value of Δx. Measure the tilt of several solar lines, estimating to a tenth millimeter. Be sure to measure at the extreme top and bottom edges of the spectra. Average your values of Δx.

(c) Determine the dispersion or scale of the spectrum for several sets of oxygen lines and obtain an average value. Measure the distances between the lines to tenths of a millimeter and use the wavelengths given in Figure 24-5.

(d) Convert your values of Δx to Angstrom units ($\Delta\lambda$) using the average dispersion.

(e) Calculate the combined velocity of the approaching and receding limbs of the sun using the Doppler equation

$$v = c \frac{\Delta\lambda}{\lambda}$$

where v = the combined velocity in km/sec,
c = the speed of light = 3×10^5 km/sec,
$\Delta\lambda$ = Doppler shift (or tilt of lines) in Å
λ = wavelength (use a line near the center of the spectrum) in Å.

(f) The rotational velocity at each limb is one-half this value because this velocity represents the difference between the two limbs. Correct for this factor of two and calculate the velocity of rotation. Remember that

$$\text{velocity} = \frac{\text{distance}}{\text{time}} = \frac{\text{circumference of sun}}{\text{period}}$$

So

$$P = \frac{2\pi R}{v}$$

Where P = the rotational period at the equator,
R = the solar radius = 6.96×10^5 km, and
v = the measured rotational velocity.

This method determines the sun's *sidereal period* or "true" period of rotation. Compare this value with that for the synodic rotational period from section 2 above.

3. ROTATION AT DIFFERENT SOLAR LATITUDES

Figures 24-7 and 24-8 were taken at solar latitudes of 60°, and 90°. They were taken with the spectrograph slit aligned across the full diameter of the sun as shown in Figure 24-2.

(a) Measure the rotational velocities for each latitude as you did in part 2, steps (a) through (e).

(b) Since these velocities are measured at higher latitudes, the small circle at each solar latitude must be used for the circumference in calculating the period. The solar radius at this latitude (r_{60}) is found by assuming the sun is a perfect sphere and using the relations shown in Figure 24-3. Find the solar rotational period at latitude 60°.

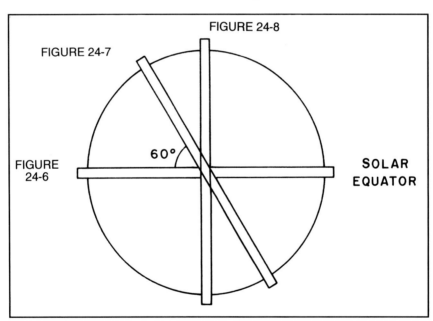

Figure 24-2. Slit Alignments for Solar Spectra.

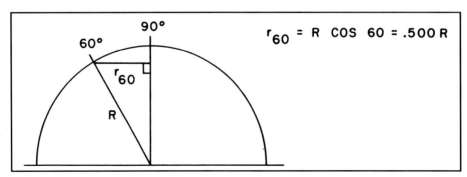

Figure 24-3. Solar Circumference as a Function of Solar Latitude.

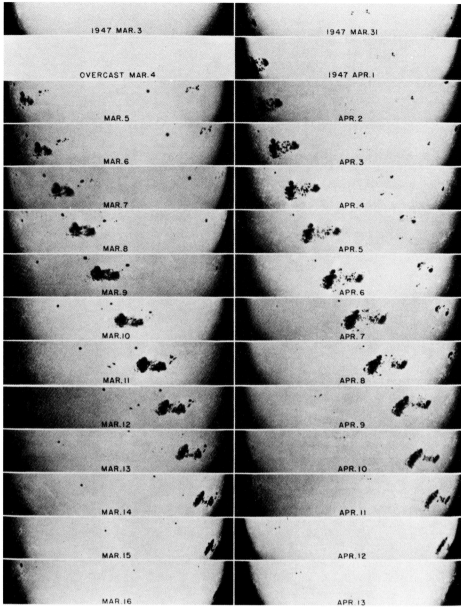

Figure 24-4. Sunspot Photographs Taken in March and April, 1947.
(Courtesy of the Hale Observatories)

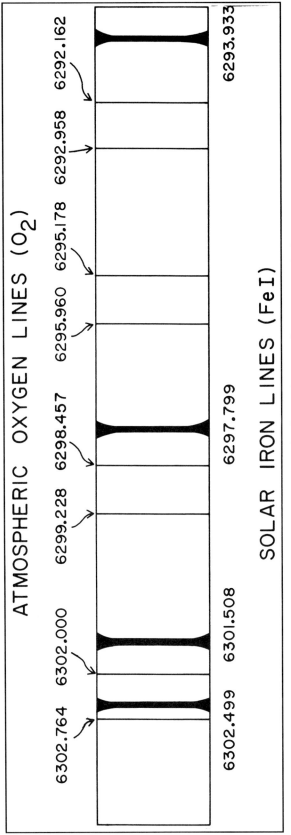

Figure 24-5. Identification of Solar and Atmospheric Spectral Lines.

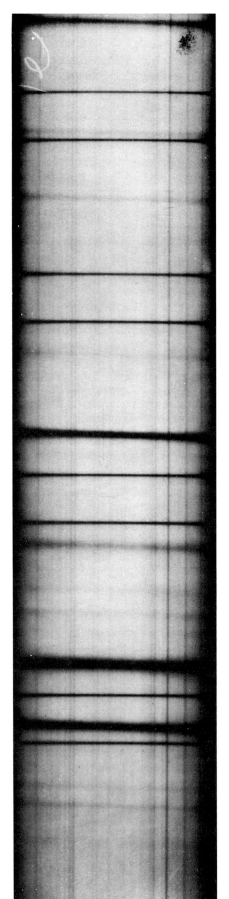

Figure 24-6. Solar Spectrum at 0 Solar Latitude.

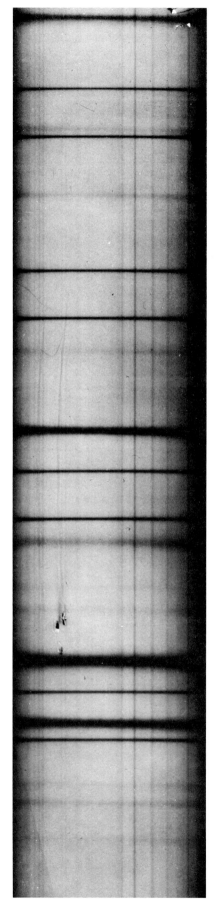

Figure 24-7. Solar Spectrum at 60 Solar Latitude.

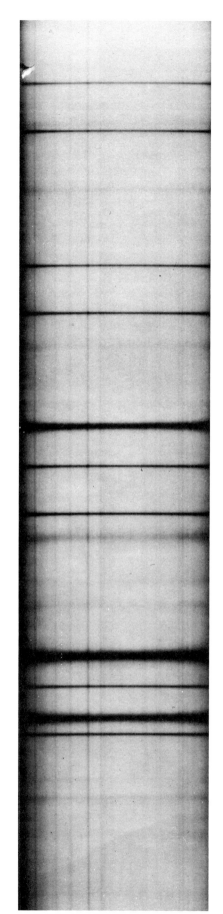

Figure 24-8. Solar Spectrum at 90 Solar Latitude.

DISCUSSION QUESTIONS

1. What changes can you see in sunspots and in sunspot groups over a time period of several weeks?

2. Estimate and compare your errors in determining the period of the sun's rotation using sunspots and spectra. What is the major source of error in each method?

3. The dispersion of the stellar and planetary spectra analyzed in this manual have typical values of a few or a few tens of angstroms per millimeter. Yet we find the solar dispersion is most easily expressed in units of mm/Å rather than Å/mm. Why don't we use such large dispersions for stellar spectra?

4. The solar spectral lines are much wider near the limbs than near the center. Can you explain this phenomenon?

5. Tabulate and compare class members' values for solar rotation at different latitudes. Discuss the results, including their implications and possible explanations in terms of physical models of the sun.

25   Proper Motion of a Star[1]

PURPOSE AND PROCESSES

The purpose of this exercise is to measure the proper motion of Barnard's star and use this proper motion, with a given parallax and radial velocity, to determine the space motion of a star. The processes stressed in this exercise include:

Designing Experiments
Interpreting Data
Using Numbers
Using Logic
Inferring

INTRODUCTION

Humankind has long studied motion in the universe: our earliest naked eye observations of the movements of the sun, moon and planets led us to build models of the solar system and gave rise to modern astronomy. With the advent of the telescope more subtle motions in the heavens became apparent. As early as 1718 Edmund Halley had noted that some stars were not fixed, but appeared to move in the sky relative to other stars. Arcturus in Boötes and Sirius in Canis Major were the first stars detected by Halley to have *proper motion*. This term is used to describe the yearly angular velocity of a star relative to a fixed field of stars, and is symbolized with the Greek letter μ. For example, Arcturus has a proper motion $\mu = 2.3''/\text{yr}$.

The star with the largest proper motion was discovered by E. E. Barnard in 1916 at Yerkes Observatory. This star, now called Barnard's Star, is a 9.5 magnitude star located in the constellation Ophiuchus. Its proper motion is so much larger than that of any other star that it is considered to be virtually a "runaway" star. Even so, the angular velocity is small enough that measurements must be made from photographs taken many years apart: the negatives of Barnard's Star in Figure 25-3 were taken in 1924 and 1951 respectively.

Since proper motion is an angular velocity, we also need to know the star's distance to find its real velocity across our line of sight, called its *tangential velocity* (v_t). Two stars with the same proper motions can have vastly different tangential velocities, as shown in the example in Figure 25-1.

Since stellar parallaxes are usually tabulated rather than stellar distances, tangential velocity can be calculated from the equation

$$v_t = \frac{4.74 \, \mu}{p} \qquad \qquad \text{Equation 1}$$

where t = tangential velocity in km/sec
μ = proper motion in seconds of arc/year
p = parallax in seconds of arc.

1. This exercise is adapted with permission from one developed by D. Scott Birney at Wellesley College.

We still need one more piece of information to know the velocity and direction a star moves in space, called its space motion or *space velocity*. So far we have considered only that part of a star's motion across our line of sight. The *radial velocity* (v_r), along our line of sight, is added vectorially (at right angles) to give us the total space velocity:

$$\vec{v} = \vec{v}_r + \vec{v}_t$$

where $\vec{v}$ = space motion

$\vec{v}_r$ = radial velocity

$\vec{v}_t$ = tangential velocity.

Remember that a negative radial velocity indicates motion toward us, and positive motion away.

The magnitude of v can be calculated using the Pythagorean theorem (Figure 25-2):

$$v^2 = v_r^2 + v_t^2$$
$$v = \sqrt{v_r^2 + v_t^2}$$

Equation 2

Space motions allow us to study the dynamics and interactions of groups of stars, and to see how the arrangement of our local group of stars is changing over the centuries.

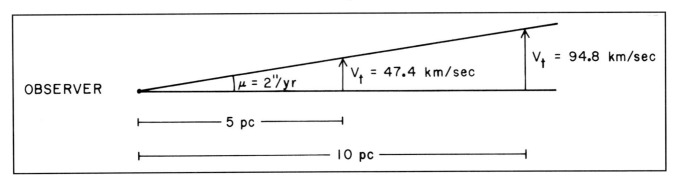

Figure 25-1. Proper Motion at Two Distances.

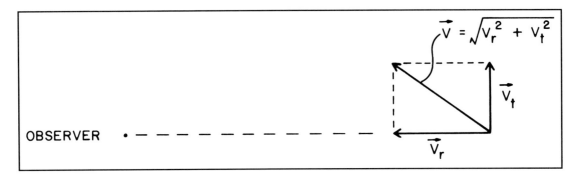

Figure 25-2. Space Velocity.

PROPER MOTION OF BARNARD'S STAR

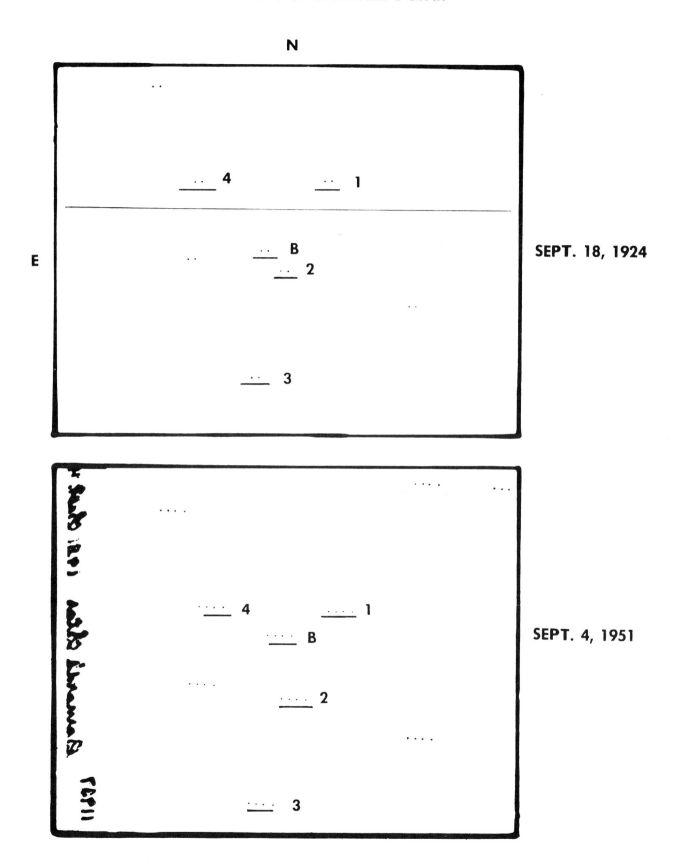

Figure 25-3. Proper Motion of Barnard's Star.

PROCEDURE
---

1. PROPER MOTION

Figure 25-3 is a negative print of Barnard's Star and four labeled reference stars taken in 1924 and 1951.

(a) Set up an arbitrary coordinate system on a piece of millimeter graph paper with the origin (x = 0, y = 0) near the lower left-hand corner. Align the graph paper over the 1924 negative so that the heavy horizontal line of the negative lays parallel to the x-axis. This establishes the east-west direction with east to the left and north toward the top of the negative. (A sheet of translucent graph paper is provided in Appendix 4.)

(b) Using the coordinate system on the graph paper trace and record the coordinates of the four reference stars and of Barnard's Star, marked "B." Note that the stars were exposed more than once; use only the right-hand image of each star.

(c) Place the graph paper over the 1951 negative and align it so the right-hand images of the four reference stars are superimposed on the plotted positions. Read and record the 1951 coordinates of Barnard's Star, and plot it on the graph paper.

(d) Measure the distance between the two positions of Barnard's Star to a tenth of a millimeter. Using a plate scale of 24.52 seconds of arc per millimeter, convert this motion to seconds of arc.

(e) These two photographs were taken exactly 26.96 years apart. Calculate the proper motion of Barnard's Star in seconds of arc per year.

2. SPACE MOTION

(a) The parallax of Barnard's Star has been measured to be 0.545″. Using Equation 1, find the tangential velocity of Barnard's Star in km/sec.

(b) Barnard's Star has a radial velocity of –108 km/sec. Calculate the magnitude of the star's space velocity by using Equation 2 or by constructing a diagram similar to Figure 25-2.

3. CLOSEST APPROACH TO THE SUN

Since Barnard's Star's negative radial velocity indicates that it is moving generally toward us, we can find the epoch of its closest approach to the sun and discover just how close that approach will be.

(a) On a piece of graph paper from a point representing the sun, draw a line to represent the direction of Barnard's Star. Calculate the distance to Barnard's Star in parsecs from the parallax-distance relation (d = 1/p). Using a convenient scale (such as 10 cm = 1 pc), mark a point at the proper distance from the sun to represent Barnard's Star.

(b) Along the line and from Barnard's Star, draw an arrow or vector representing the radial velocity of the star (a scale of ½ mm = 1 km/sec works well). Again from Barnard's Star draw another vector (at right angles) representing the tangential velocity, using the same scale as for v_r. Complete the rectangle as in Figure 25-2 to find the direction and magnitude of the space velocity of Barnard's Star.

(c) To find the point of Barnard's Star's closest approach to the sun extend the diagonal of the space velocity to show the path the star will take. Using a compass or ruler, find the point along this line closest to the sun. A line connecting the sun and point of closest approach should be perpendicular to the path of Barnard's Star. Measure the closest approach distance in cm and convert to parsecs using the same scale factor as in part 3a. How close will Barnard's Star come to the sun?

(d) By again measuring from your diagram, determine how far (in pc) Barnard's Star has to travel from its present position to its point closest to the sun. From the velocity and distance it has to travel we can calculate how long it will take to get nearest to us: velocity = distance/time, so time = distance/velocity. Using the conversion factor 1 km/sec = 1.02×10^{-4} pc/century, when will Barnard's Star be closest to the sun?

DISCUSSION QUESTIONS

1. At present our closest star is the Alpha Centauri system at about 1.33 pc. How will Barnard's Star compare to this at its closest approach?

2. How do we know that the four reference stars don't have proper motions of their own?

26 Spectral Classification

PURPOSE AND PROCESSES

The purpose of this exercise is to become familiar with the classification system of stellar spectra. The processes stressed in this exercise include:

Observing
Classifying
Inferring

INTRODUCTION

The discovery of the underlying causes of stellar spectra completely changed the history of astronomy. Auguste Comte defined *astronomy* in 1835 strictly in terms of position and movements of heavenly bodies and completely ruled out the possibility that we would ever know the compositions of these objects.

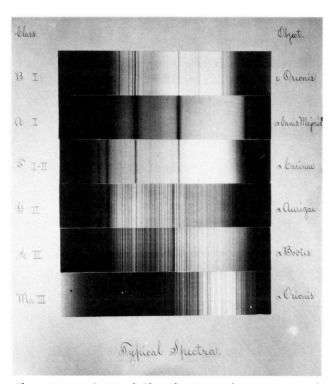

Figure 26-1. Original Classification Scheme Proposed by Harvard College Observatory in 1901. (Courtesy of Harvard College Observatory)

By the 1860s Kirchoff and Bunsen had laid the groundwork for the development of the science of spectroscopy. Building on this initial work, astronomers such as Secchi and Pickering developed classification schemes for categorizing the large numbers of stellar spectra. It was, ultimately, the work of Anne Jump Cannon and co-workers at the Harvard College Observatory that led to our modern classification system. Even then, investigators believed the different appearances of stellar spectra resulted from differences in stellar chemical compositions. Later, another Harvard astronomer, Cecilia Payne-Gaposchkin demonstrated that the differences in spectra reflected temperature differences in the stars and that the chemical composition of stars are remarkably similar. Hence, an examination of stellar spectra made possible the determination of the composition and temperatures of stars and subsequently led to an expansion of knowledge and the development of a separate branch of science, astrophysics.

The most extensive spectrum classification project in the world was completed at Harvard in the first quarter of this century. As the result of a large bequest from the estate of a New York physician, women astronomers were hired and ultimately completed the classification of approximately a quarter million stars. This famed Henry Draper Catalogue, named in honor of its benefactor, set the standard for stellar classification that remains until this day. The classification system it is based upon makes use of just a few categories and is known as the Draper system or the Harvard classification system.

While the system has been modified and extended over the past three-quarter century, the system has proven to be very durable. Figure 26-1 is a copy of the original classification spectra first published at Harvard in 1901. It is remarkably similar to reference systems used today. The spectra shown are absorption spectra. This means that what we are seeing is the result of the energy from the photosphere of the star which passed through a cooler atmosphere. The stellar atmosphere absorbed some of the energy at certain wavelengths and left narrow regions of the spectra with less energy, hence the darker appearance of the spectral lines.

The Harvard sequence consists of seven classes which are titled with the rather confused sequence of letters O-B-A-F-G-K-M. This sequence is the result of earlier work in which stars were assigned to category designations which started with the letters of the alphabet. Later, examination revealed that some categories could be lumped together and re-arranged in a more systematic manner. The sequence is arranged in order of decreasing temperature. The O stars are the hottest stars and the M stars are the coolest stars. Decimal subdivisions were added so that a star that was intermediate between a B star and an A star would become known as an A5 star. The Sun is a G2 star. The majority of stars fit the sequence outlined above. There are a few stars which require special classification. Types R and N are sub-classes of K and M stars respectively. Wolf-Rayet stars are O type stars with broad emission lines which are thought to originate from materials ejected from the star. This ejected gas absorbs light from the star and re-emits it as emission lines.

PROCEDURE

---

1. Carefully examine the spectral key given in Figure 26-3 and compare the spectra with the following descriptions.

   O-stars. Stars in this category are extremely hot, greater than 25,000 K. The relatively few absorption lines are those of ionized helium and a few weak doubly ionized nitrogen, triply ionized silicon and hydrogen lines.

   B-stars have temperature ranges between 11,000 and 25,000 K. The lines present are those of neutral helium, and singly and doubly ionized silicon and singly ionized oxygen and magnesium. Hydrogen lines are more pronounced than in O-type spectra.

   A-stars have very strong hydrogen lines as well as lines of magnesium, silicon, iron, titanium and calcium. Their temperatures range from 7,500 to 11,000 K.

   F-stars have surface temperatures in the 6,000 to 7,500 K range. Hydrogen lines are weaker but still obvious. Lines of singly ionized calcium, iron and chromium are present. Some lines of neutral metals are seen.

   G-stars are cooler still, 5,000–6,000 K. Lines of ionized calcium are the most conspicuous spectral feature. Many lines of ionized and neutral metals are present. Hydrogen lines are weaker than in F stars.

   K-stars are quite cool by astronomical standards, 3,500–5,000 K, and lines of neutral metals predominate in the spectra. Some molecular bands appears.

   M-stars are quite red and are cooler than 3,500 K. Strong lines of neutral metals and molecular bands dominate the spectra.

   WN-stars have broad emission features. The N refers to the presence of nitrogen lines.

   The spectra shown in this exercise were taken with film with a blue-sensitive emulsion. As such the spectra shown cover only about 4000 to 5500 Angstroms. Also notice that the shorter wavelengths are shown on the left end of the spectra and the longer wavelengths are on the right.

2. Figure 26-2 is a plate with an objective-prism spectrogram taken on the evening of October 16–17, 1950 at the Warner and Swasey Observatory. A large, thin prism was fitted to the front of a 24-inch Schmidt telescope and a four-minute exposure was made of the region of the sky near the star, Gamma Cygni. The field of view is approximately 5 degrees. The brightest star in the field of view is the 5th magnitude P Cygni near the right (north) edge of the photograph, but its spectrum is over-exposed. The best exposed spectra on the plate are from stars of the 7th to 9th magnitude. In order to make the spectra easier to

examine, the images were trailed in an east-west direction. The effect of this trailing is to broaden the spectra and this makes their details easier to examine.

Carefully examine the plate. You should be able to see the diversity of spectra present. The region of the sky chosen for the plate shows examples of nearly every Harvard spectral type.

3. Working with a laboratory partner, look at the first five numbered spectra. Compare these spectra with the reference spectra shown in Figure 26-3. Observe not only the spacing of the lines present, but the relative intensity of the lines. Determine what spectral type you believe these five examples are. Record your answers. After a few minutes, share your answers with your instructor. Your instructor will then supply the spectral types agreed upon by professional astronomers. Using this feedback should assist you in classifying the rest of the numbered spectra on the plate.

4. Examine the remaining numbered spectra from Figure 26-2. Using the reference spectra, assign spectral types to each of the stars. Remember to assign decimal subdivisions as needed. Record your answers.

Figure 26-2. Objective Prism Spectrogram.
(Courtesy Warner and Swasey Observatory)

4. Spectral Classification-Data Sheet

1. My estimates for the spectral type of stars 1–5 are:

    1. _____

    2. _____

    3. _____

    4. _____

    5. _____

2. My estimates for the spectral type of stars 6–25 are:

    6. _____

    7. _____

    8. _____

    9. _____

    10. _____

    11. _____

    12. _____

13. _____

14. _____

15. _____

16. _____

17. _____

18. _____

19. _____

20. _____

21. _____

22. _____

23. _____

24. _____

25. _____

Figure 26-3. Reference Spectra.

DISCUSSION QUESTION

Notice that the spectra of hotter stars are brighter on the left end (shorter wavelengths) than on the right side. Notice that the opposite is true of the cooler stars. What explanation can you give for this?

27

A Color-Magnitude Diagram of the Pleiades[1]

PURPOSE AND PROCESSES

The purpose of this exercise is to construct and examine a color-magnitude diagram of the Pleiades star cluster. The processes stressed in this exercise include:

Using Numbers
Interpreting Data
Formulation of Models

INTRODUCTION

As astronomers developed techniques for studying some of the physical properties of the stars, they also searched for relationships among these properties. Two of the most easily measured properties of any star are its apparent magnitude, measured either photographically or photoelectrically, and its temperature, determined from its stellar spectral class or *color index*.

Color index is a comparison of the amounts of blue and yellow (or visual) light a star radiates: a hot star radiates more blue and violet light, while a cooler one radiates more yellow and red light (Figure 27-1).

Color index can be defined then as

$$CI = m_b - m_v$$

and is often written

$$CI = B - V$$

where CI is the color index
$m_B = B$ is the apparent blue magnitude
$m_v = V$ is the apparent visual magnitude.

Apparent stellar magnitudes in the blue spectral region can easily be measured photographically because most film is more sensitive to blue light. Visual magnitudes are measured from photographs taken with an amber filter. Standardized blue, visual, and ultraviolet filters have been developed for photoelectric use.

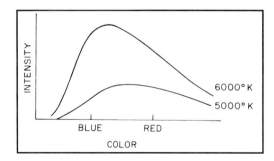

Figure 27-1. Radiation curves.

1. This exercise is adapted with permission from Michael K. Gainer, ASTRONOMY: OBSERVATIONAL ACTIVITIES AND EXPERIMENTS. Copyright © 1974 by Allyn and Bacon, Inc., Boston.

In using the color index of a star we must remember the way the stellar magnitude scale is set up: large magnitude numbers refer to dimmer stars. Thus, for a hot blue star, B will be numerically less than V (indicating more blue than visual light) and the color index will be a negative number. Likewise, cooler stars have positive color indices.

Because a star's apparent magnitude depends on its distance as well as its intrinsic brightness, we really would not expect a correlation between apparent magnitude and temperature (or color index) for most stars in the sky. However, it might seem reasonable for a star's temperature to be related to the total amount of energy it radiates, or its absolute magnitude. Unfortunately, absolute magnitudes are hard to determine directly because of the difficulty in determining reliable distances for more than a few hundred stars in the sky.

One way to avoid this difficulty is to study the magnitudes and colors of stars in a cluster. All stars in the cluster are at approximately the same distance, so the apparent magnitude of each differs from its absolute magnitude by the same factor. By comparing color index and apparent magnitude we are also relating temperature and absolute magnitude.

PROCEDURE

Figures 27-2 and 27-3 are photographs of the Pleiades taken without and with an amber filter, respectively. A 135 mm focal length lens was used with Tri-X film at f/2.8, with exposure times of 1.0 and 1.5 minutes. (A longer exposure time must be used with a filter to compensate for the absorption of light by the filter.)

1. Try to obtain your own photographs of the Pleiades if possible. Lenses with focal lengths from 100 mm to 200 mm work well at the lowest f-ratio setting, with exposure times ranging from one to three or four minutes. Allow approximately one and one-half times as long an exposure time for the photograph with the filter. To minimize the effects of possibly changing atmospheric conditions, alternate frames taken with and without the filter rather than taking a series with the filter followed by a series taken without.

2. Figure 27-4 is a comparison scale to use in estimating stellar brightnesses. Number the images from 1 to 9, starting with the largest. Although these numbers do not correspond to the standard stellar magnitude scale, they are proportional to magnitudes. Cut out the comparison scale so that it can be placed adjacent to the star images in Figures 27-2 and 27-3.

3. Measure the magnitude of a star on the print for which no filter was used by comparing the size of the star image with the comparison scale. If a star image falls between two images on the scale, say between 4 and 5, assign it a value of 4.5.

   After you measure a particular star on the print for which no filter was used, move immediately to the print obtained with the amber filter and measure the same star. Number each star after you measure it to avoid measuring the same star twice.

4. Measure at least 20 stars, selecting samples of all magnitudes. Select stars only from around the center of the print. Do not measure stars which cannot be distinguished as single stars.

5. Plot a color-magnitude diagram of B (from the photograph taken without a filter) versus B–V, the difference between the two magnitudes.

6. Discuss any conclusions you can make from this graph.

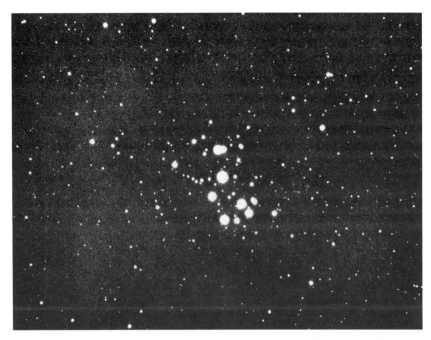

Figure 27-2. A Photograph of the Pleiades Taken Without a Filter.

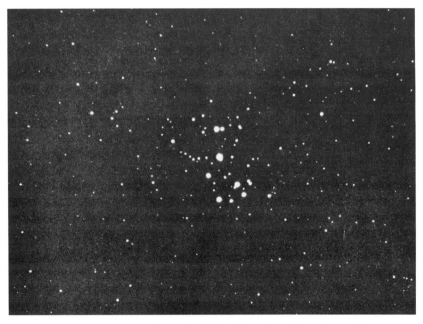

Figure 27-3. A Photograph of the Pleiades Taken Through an
85B Amber Filter.
(Courtesy of Michael K. Gainer, St. Vincent College)

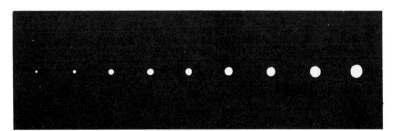

Figure 27.4. Magnitude Comparison Scale.

27. A Color-Magnitude Diagram of the Pleiades

B

B - V

DISCUSSION QUESTION

1. How do you interpret data that fall well outside the majority of the points on your plot?

28

Distance to the Pleiades

PURPOSE AND PROCESSES

The purpose of this exercise is to plot a color-magnitude diagram for the Pleiades star cluster and to determine its distance. The processes stressed in this exercise include:

Using Numbers
Interpreting Data
Controlling Variables
Inferring
Predicting

INTRODUCTION

An open star cluster consists of a few hundred stars of common age and origin, loosely held together by mutual gravitation, and moving together through space. They usually are located in the main disc of a galaxy, in or near the spiral arms. Open clusters are therefore often called galactic clusters (not to be confused with clusters of galaxies).

Open clusters do not contain RR Lyrae stars, and only a few contain cepheid variables. In addition, distances cannot be determined from angular diameters because open clusters differ from one another in size and their angular diameters do not depend on distance alone. Distances to open clusters must therefore be determined by means different from those used for globular clusters. In practice, the most useful method for distance determination is to use the color-magnitude diagram.

Figure 28-1 shows a negative print of a photographic plate of the Pleiades, taken with a 400 mm lens and f/6.3 for 20 minutes. Its distance can be found by first plotting a color-magnitude diagram and then fitting the data from this graph to a standard Hertzsprung-Russell curve. The approximate diameter of the cluster can also be found by counting the number of stars in concentric equal-area rings.

PROCEDURE

Table 28-I lists colors and apparent magnitudes for many of the stars in the Pleiades cluster. The Eggen number refers to a system for identifying Pleiades stars, m_b is the blue or photographic magnitude, m_b is the visual magnitude, and $m_b - m_v$ is the color index. This information is similar to that obtained by Hertzsprung and Russell for stars of the near solar neighborhood, and also to that obtained in Exercise 27 "A Color-Magnitude Diagram of the Pleiades." The important difference between these data and those in the previous exercise is that here the magnitudes are given in standard magnitude units. The arbitrary magnitude scale of Exercise 27 is useful for comparing the stars within the cluster itself; however, they do NOT allow comparison with the standard main sequence to determine distance.

1. Plot a color-magnitude diagram for the Pleiades. Watch the signs on the $m_b - m_v$ values.

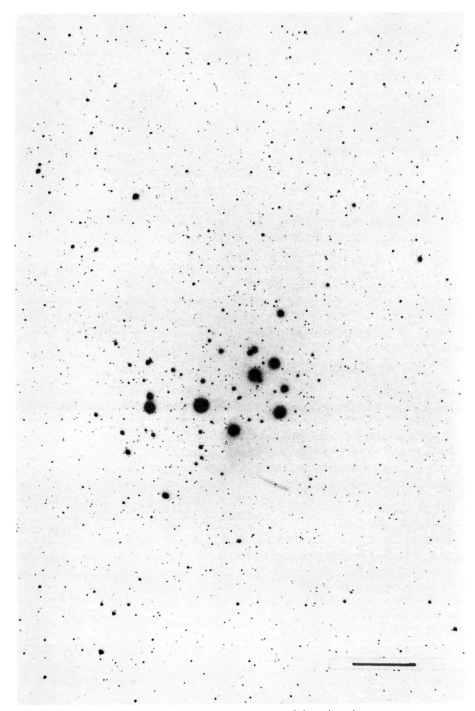

Figure 28-1. A Negative Print of the Pleiades.
(Courtesy Dean Ketelsen, University of Iowa Observatory)

2. Your diagram may reveal only a portion of the main sequence. The reason for the lack of a complete main sequence in many clusters is two-fold. The curve is often branched on the upper part of the main sequence as a result of the aging process of the stars in the cluster. The curve is often also truncated on the lower right for a different reason. Why?

3. For most clusters there is still enough of a main sequence on the diagram to provide the information needed to find the cluster's distance. On a second graph, USING THE SAME SCALE, plot an HR diagram for the stars given in Table 28-II. These are stars for which distances have been determined by independent means, so that the absolute magnitudes have been calculated. Note that here the absolute visual magnitude (M_V) and the $M_B - M_V$ color index are given.

Table 28-I
Colors and Magnitudes for the Pleiades[1]

Eggen Number	$m_b - m_v$	m_v	Eggen Number	$m_b - m_v$	m_v
3	+0.085	8.24	26	+0.038	7.85
5	+0.043	8.06	27	−0.215	5.74
7	+0.332	9.60	28	−0.197	6.41
8	+0.118	8.14	31	−0.242	4.17
9	+0.414	9.83	32	+0.526	10.42
10	−0.224	5.44	33	−0.073	7.34
11	−0.297	3.70	34	+0.209	8.09
13	+0.512	10.37	35	+0.620	10.20
15	+0.197	8.56	36	+0.343	9.27
16	−0.255	5.64	39	+0.527	10.51
17	−0.289	4.29	40	−0.148	6.81
18	+0.307	8.97	41	+0.149	8.37
19	+0.067	8.03	42	+0.339	9.44
20	+0.211	8.58	43	−0.140	6.98
21	+0.487	10.11	44	+0.046	7.64
22	+0.006	7.15	45	−0.133	7.25
23	+0.425	9.70	46	−0.001	7.75
24	+0.369	9.42	47	−0.098	6.80
25	−0.249	3.86	56	−0.031	6.93

1. Adapted from H. L. Johnson and W. W. Morgan, Ap. J. 114, 522 (1951).

Table 28-II
Standard Main Sequence

The location of the Main Sequence has been studied by numerous authors in order to make an assignment of colors, spectral types and absolute magnitudes. The following list is a tabulation from several authors as compiled by Evans in his book *Observations in Modern Astronomy.*

$M_B - M_V$	M_V	$M_U - M_B$	Type	$Temp_{(eff)}$
−0.25	−2.10	−0.89	B2 V	20,300
−0.20	−1.10	−0.70	B3 V	18,000
−0.15	−0.30	−0.52	B5 V	15,000
−0.10	0.50	−0.33	B8 V	12,800
−0.05	1.10	−0.16	B9 V	11,800
0.00	1.50	0.00	A0 V	11,000
0.05	1.74	0.05	A2 V	9,700
0.10	2.00	0.08	A3 V	9,100
0.20	2.45	0.07	A7 V	8,100
0.30	2.95	0.02	F0 V	7,000
0.40	3.56	−0.01	F3 V	6,800
0.50	4.23	0.00	F7 V	6,300
0.60	4.79	0.08	G0 V	6,000
0.70	5.38	0.24	G8 V	5,320
0.80	5.88	0.43	K0 V	5,120
0.90	6.32	0.63	K2 V	4,760
1.00	6.78	0.87	K3 V	4,610
1.10	7.20	1.03	K4 V	4,500
1.20	7.66	1.13	K5 V	4,400
1.30	8.11	1.21	K6 V	4,000

From D. S. Evans, *Observations in Modern Astronomy,* p. 95, 1968; by permission of the English Universities Press Limited.

4. At least two different methods can be used to determine corresponding values of M_V and m_v from your data. Try using both methods and compare your results.

(a) Read the apparent magnitude for a star of a given $m_b - m_v$ from one graph, and the corresponding absolute magnitude of a star of the same color from the second graph. Repeat for several other stars in the cluster, and calculate an average value of $m_v - M_V$, called the *distance modulus*, and often written simply m − M.

(b) Place the color-magnitude diagram over the standard main sequence plot, aligning the two color indices. Keeping these scales aligned, slide the top graph up and down until the color-magnitude data points best overlay the standard main sequence curve. Read corresponding values of M_V and m_v from the vertical axes of the graphs, and find the distance modulus.

5. Calculate the distance to the Pleiades using the equation:

$$\log d = \frac{m - M + 5}{5}$$

where d = distance in parsecs
     m = apparent magnitude
     M = absolute magnitude.

Optional Activity

This optional section can be done in two ways. Choice A is likely to produce better results but is quite time consuming. Choice B is faster yet still produces good results.

A. You can calculate the diameter of a cluster in parsecs if you measure its angular diameter and know its distance, from the relation

$$s = r\theta$$

where s = linear size (pc)
     r = distance (pc)
     θ = angular size (radians).

(You might want to review this relation in Exercise 10 "Angles and Parallax," and remember that 1 radian = 57.3°.)

1. A good method of determining the angular size of a cluster is to construct concentric equal-area rings around the estimated cluster center. Count the number of stars in successive rings and estimate the size from where this count falls sharply. Determine the effective size of the cluster in cm.

2. The scale of the photograph is indicated by the heavy line in the lower right-hand corner, with its length representing 30 minutes of arc. Determine the scale of the photograph in cm/arc min, and find the angular size of the cluster.

3. Calculate the cluster diameter in parsecs.

4. Assuming the cluster is roughly spherical, estimate its volume in cubic parsecs.

5. The density of stars in the solar neighborhood is approximately one star per 10 cubic parsecs. How does the density of stars in the cluster compare to the density of stars near us?

B. Using a drawing compass place the point at what appears to be the center of the cluster. Expand the compas until it appears that you have all the bright stars covered with a circle constructed out to the radial distance. Then determine the size of the cluster in cm. Proceed as is steps 2–5 in Option A.

28. Distance to the Pleiades

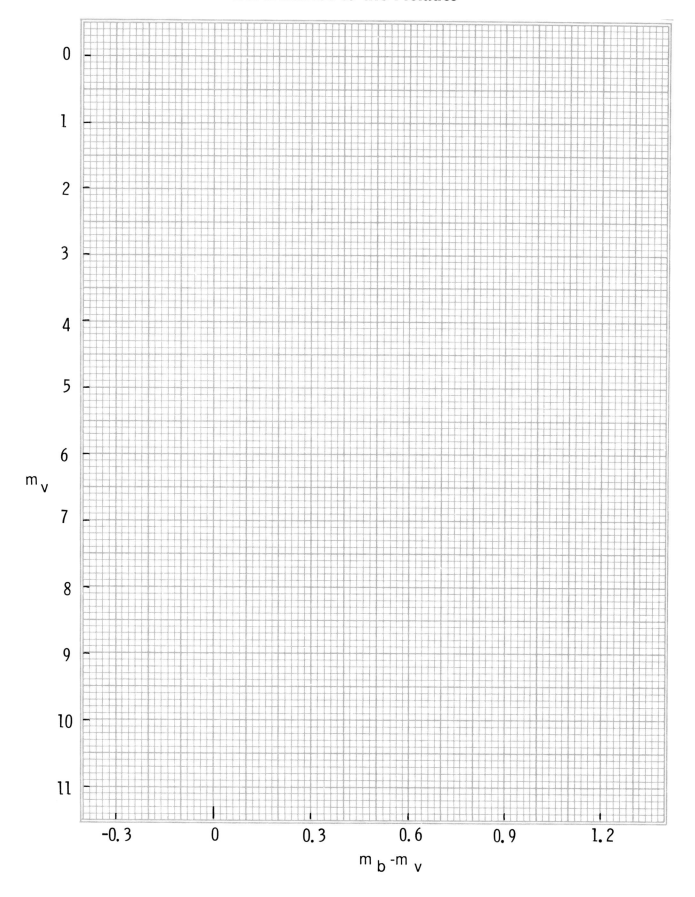

$m_b - m_v$

28. Distance to the Pleiades

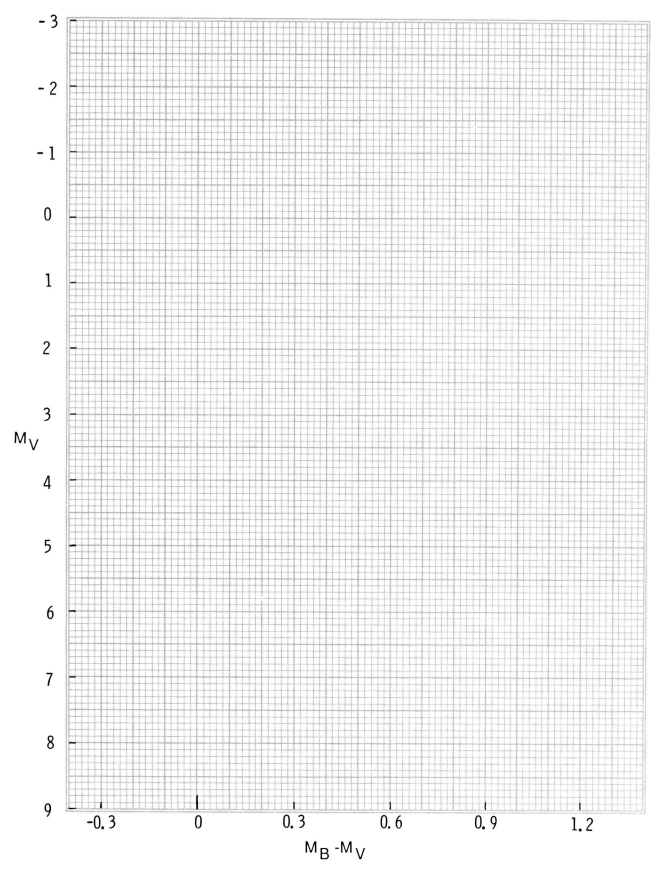

$M_B - M_V$

M_V

DISCUSSION QUESTIONS

1. Which method of determining corresponding values of m_v and M_V do you think was most accurate? Estimate your uncertainty in magnitude units for each method. Would you expect the two methods to be equally valid for all clusters?

2. What assumptions are made in determining the distance to a cluster using these methods?

3. How would interstellar reddening affect your results? Would all stars on your color-magnitude diagram be equally affected?

4. Where might this cluster be located from its distance and from your knowledge of the properties of open clusters?

29   Galactic Clusters and HR Diagrams[1]

<table>
<tr><td>PURPOSE AND PROCESSES</td></tr>
</table>

The purpose of this exercise is to study the relative HR diagrams (or color-magnitude diagrams) of several open clusters to determine their ages and distances, and to detect the effect of interstellar reddening on cluster observations. The processes stressed in this exercise include:

> Questioning
> Interpreting Data
> Using Numbers
> Formulation of Models
> Inferring
> Using Logic
> Prediction

INTRODUCTION

In the previous exercises on the Pleiades we have seen that a color-magnitude or HR diagram of a star cluster allows us to study stellar properties in general, and to determine cluster properties such as distance. The study of a number of galactic clusters also has produced some very important information about stellar evolution. The *zero age main sequence* (ZAMS) is the locus of points on the HR diagram where stars of different masses first begin to generate all their energy by the fusion of hydrogen.

The ZAMS therefore can be defined by the color-magnitude diagrams of some very young open clusters. This process also can be reversed to determine the ages and distances of many other galactic clusters. This method compares a cluster's color-magnitude diagram to the ZAMS on a standard HR diagram.

Figures 29-1 through 29-9[2] show the color-magnitude diagrams for nine open clusters. Note that the vertical axes give the apparent visual magnitude (V) of the stars in the clusters rather than the absolute magnitude (M_v) found in most HR diagrams. Color-magnitude diagrams are sometimes called "relative" HR diagrams because the apparent magnitude depends on distance as well as the intrinsic brightness of stars.

The top axis gives $(B-V)_O$, the intrinsic color index (indicating spectral type) of each star in the cluster. The bottom axis gives the color index (B-V) for each star as actually measured. The reason for the difference in these scales will be examined later in this exercise.

Figure 29-10 (in Appendix 4) is a translucent overlay showing the ZAMS on an HR diagram. Note that here the axes are labeled absolute magnitude and intrinsic color.

---

1. This exercise is adapted with permission from one developed by John D. Fix at The University of Iowa.
2. Cluster diagrams and ZAMS overlay are from Gretchen L. Hagen, Publications of the David Dunlap Observatory, 4 (1970); by permission of the author.

PROCEDURE

1. CLUSTER AGE

We can assume that all stars in a given cluster have about the same age and original chemical composition. All stars of a cluster don't fall on the main sequence, however, because stars of different masses evolve at different rates. In addition to main sequence stars, a cluster may contain stars that are just evolving away from the main sequence, such as red giants. In general, we might expect that in an older cluster more stars would have had time to evolve from the main sequence. From studies of stellar evolution we find that this is just the case: the more massive O and B type stars leave the main sequence first. By finding the *turn-off point*, that region of the main sequence where the stars are just beginning to evolve toward the red giant region, we can estimate the cluster's age. Figure 29-11 shows a graph of $(B-V)_O$ of this turn-off point versus the age for clusters.

(a) Place the transparency over the HR diagram of each cluster and slide it left or right until the $(B-V)_O$ scales for the two diagrams match. Now slide the overlay up and down (keeping the $(B-V)_o$ scales aligned) until you get the best possible match between the ZAMS and the main sequence for each cluster. Remembering that hot main sequence stars (having small or negative color indices) leave the main sequence first, be sure that the cooler main sequence stars fall on the ZAMS even if the hotter stars do not.

(b) After matching the ZAMS and the cluster main sequence, find the turn-off point of the main sequence for the cluster. This point should be carefully chosen so that there are few stars lying to the left of it. Use Figure 29-11 to determine the age of each cluster.

2. CLUSTER DISTANCE

The M_V scale of the HR diagram transparency and a cluster's V scale will match only if the cluster is ten parsecs away because a star's absolute magnitude is the apparent magnitude it would have if it were at a distance of ten parsecs. In fact, none of these clusters is as close as ten parsecs, so the apparent magnitudes are bigger (indicating the stars are dimmer) than the corresponding absolute magnitudes on the ZAMS overlay. By comparing the corresponding V and M_V we can determine cluster distances in parsecs using the relation

$$\log d = \frac{V - M_V + 5}{5}$$

where d = cluster distance in pc,
  V = apparent visual magnitude
  M_V = absolute visual magnitude.

Matching the cluster and ZAMS $(B-V)_O$ scales as in part 1, read the corresponding values of M_V and V. Find $V - M_V$ (the *distance modulus*) for each cluster and determine the distance to each using either the equation above or the graph of distance modulus vs. distance in Figure 29-12.

3. COLOR EXCESS AND REDDENING

As noted earlier, the (B-V) and $(B-V)_O$ scales of the cluster diagrams do not line up: in all cases the lower (B-V) scale is shifted to the left of the upper $(B-V)_O$ scale. For example, we might measure the (B-V) for a star to be 0.4, yet on the upper scale we see its intrinsic color index is 0.3. We are observing the stars in a cluster to be redder than they really are because larger positive numbers refer to redder stars.

This difference between apparent or measured color index and the intrinsic color index is called the *color excess* of a cluster and can be written

$$CE_{B-V} = (B-V) - (B-V)_O.$$

(a) Measure the color excess for each of the nine clusters.

(b) Table 29-I gives the actual distances to the nine clusters. How do your distance determinations compare to the real distances? How can you account for the discrepancies (if any) which exist?

(c) Calculate the ratio of the distance you found to the actual distance of each cluster. Make a graph showing this ratio versus CE_{B-V}. Is this ratio related to the color excess of a cluster? If so, how do you account for this fact?

Table 29-I
Cluster Distance

Cluster		Distance
NGC	457	2900 parsecs
NGC	752	360
Mel	20	170
M	45	120
NGC	2632	160
NGC	2682	800
IC	4725	600
NGC	6705	1700
NGC	6791	5200

DISCUSSION QUESTIONS

1. Would you expect to see many clusters younger than 10^7 years or older than 5×10^9 years? Why or why not?

2. What data from stars might we use to determine if they are reddened appreciably?

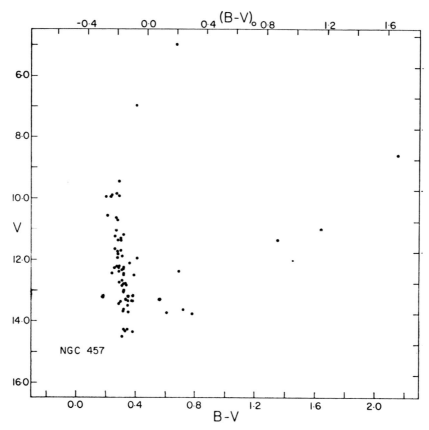

Figure 29-1. A Color-Magnitude Diagram of NGC 457.

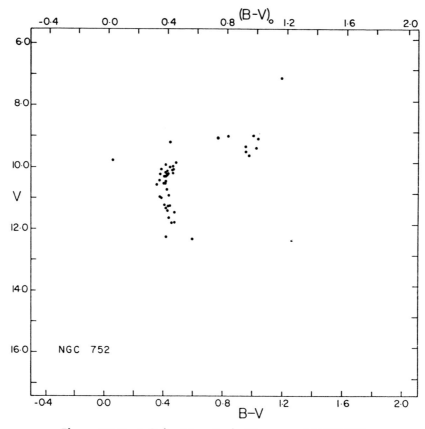

Figure 29-2. A Color-Magnitude Diagram of NGC 752.

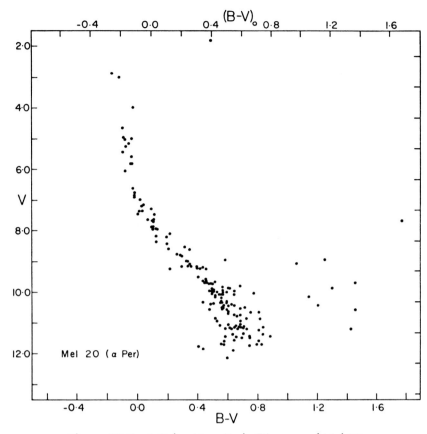

Figure 29-3. A Color-Magnitude Diagram of Mel 20.

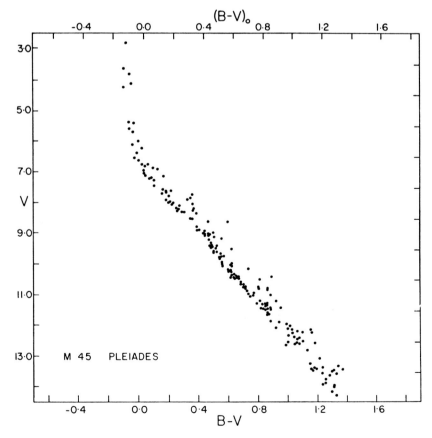

Figure 29-4. A Color-Magnitude Diagram of M 45.

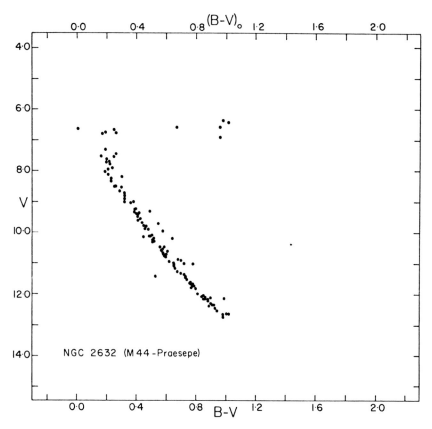

Figure 29-5. A Color-Magnitude Diagram of NGC 2632.

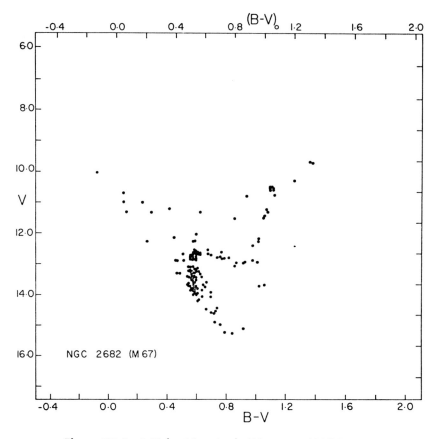

Figure 29-6. A Color-Magnitude Diagram of NGC 2682.

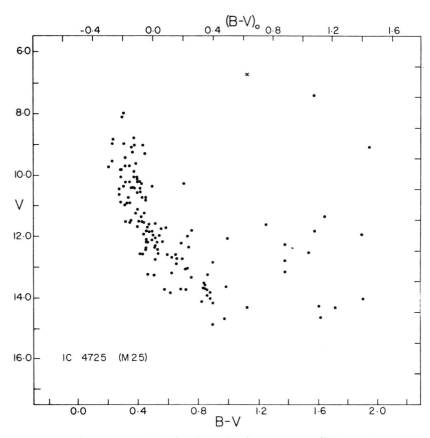

Figure 29-7. A Color-Magnitude Diagram of IC 4725.

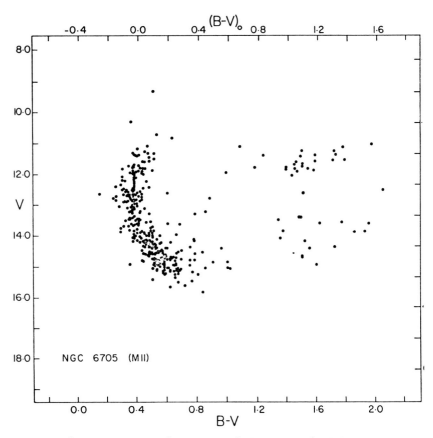

Figure 29-8. A Color-Magnitude Diagram of NGC 6705.

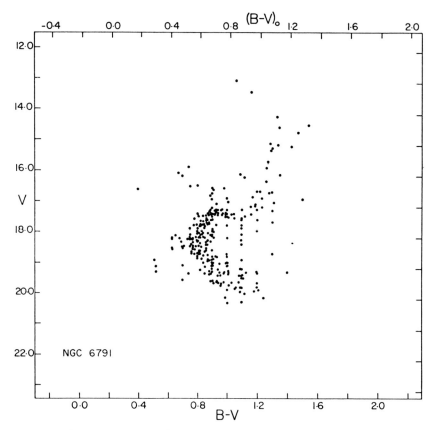

Figure 29-9. A Color-Magnitude Diagram of NGC 6792.

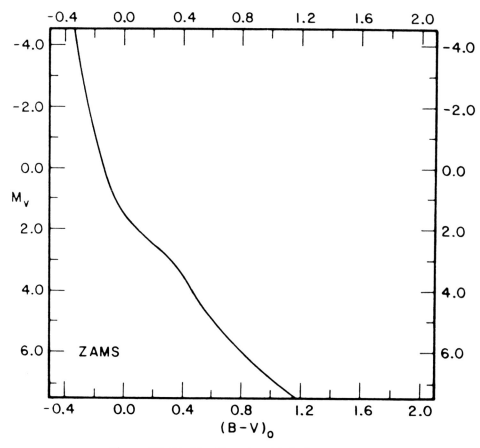

Figure 29-10. The Zero Age Main Sequence.

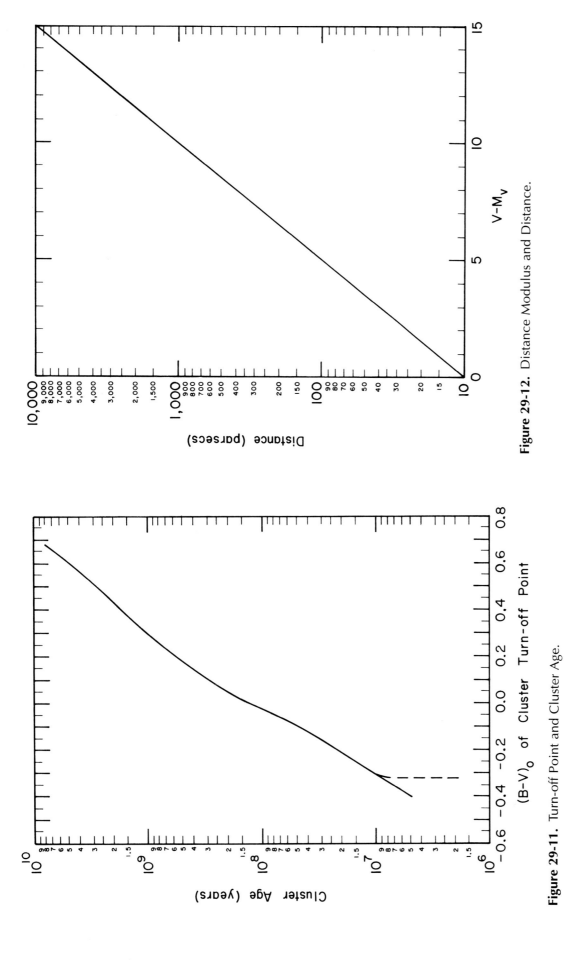

Figure 29-12. Distance Modulus and Distance.

Figure 29-11. Turn-off Point and Cluster Age.

30

The purpose of this exercise is to determine the absolute magnitude of a supernova when it is at peak brightness using measured image diameters on photographic plates. The processes stressed in this exercise include:

Using Numbers
Interpreting Data
Inferring
Controlling Variables

INTRODUCTION

Supernova are spectacular explosive events that happen in the dying stage of massive stars. Such events have been noted by astronomers for many centuries, but the mechanism by which the star actually explodes is not completely understood. Modern astronomers are very interested in supernovae, but such events are relatively rare. Astronomers estimate that in a typical galaxy there may be no more than one supernova per century and one has not been observed in the Milky Way for nearly 400 years!

Supernovae are commonly divided into two classes by astronomers, Type I and Type II. Type I supernovae brighten to a peak magnitude of about –19 in just a few days. They fade very quickly, for a few days, then more slowly over a period of a year or more. Type II supernovae are not as bright at peak brightness and they do not fade in as regular a fashion as Type I supernovae. The differences between the two types is thought to be caused by the differences in the pre-explosion stars. Type I supernovae are believed to occur in binary systems in which a star loses material onto the surface of its white-dwarf companion star. The resulting extra load causes the white dwarf to collapse and produces the supernova explosion. A Type II supernova occurs when a massive single star collapses after using up its available hydrogen fuel.

The Large Magellanic Cloud (LMC), a small irregular galaxy, is a satellite of the Milky Way and is located about 50 kiloparsecs away. Its name attests to the fact that it was discovered by Magellan on his historic voyage around the world. On February 23, 1987 astronomers witnessed a naked-eye supernova that came to be known as Supernovae 1987 A (SN1987 A). The name denotes that it was the first supernova observed that year. The occurrence of a supernova this close to Earth ranks as one of the most exciting astronomical events in recent years.

The placement and timing of SN1987 A was very fortuitous. It was located in a galaxy whose distance is accurately known. It was observed almost from the beginning of the explosion and (unlike all previous supernovae) information was known about the progenitor star. One disadvantage was that the event could only be seen from southerly latitudes. The United Kingdom 1.2 m Schmidt Telescope, located at Coonabarabran, New South Wales, Australia produced an excellent series of photographs showing the changing brightness of the star. A portion of that photographic series is shown on the following pages, figures 30-2, 30-3, and 30-4. These negative prints also show the bright nebula, 30 Doradus, up and to the left of the supernova. Table 30-I lists the dates on which the photographs were taken.

Table 30-1
Photograph Dates

Photograph	Date	Days Since Event
a	February 27, 1987	4
b	March 10	15
c	March 29	34
d	April 8	44
e	April 24	60
f	May 5	71
g	May 25	90
h	July 4	131
i	August 15	173
j	October 17	236
k	December 9	289
l	February 11, 1988	353

As SN1987 A changed in brightness, the diameter of the image of the star changed on photographic plates. This change in image size is proportional to the change in the *brightness* of the star and should not be interpreted as a change in the diameter of the star itself. (See Exercise 27, "A Color-Magnitude Diagram of the Pleiades" for another example of the use of photographic photometry.) The diameter of stellar images can be measured and a light curve constructed. A light curve consists of a plot of the brightness of an astronomical object plotted against the time at which each measurement was made. Light curves reveal information in a graphical form that assist astronomers in interpreting the behavior of variable objects. You will use the light curve to determine the apparent magnitude of SN1987 A at its brightest. From that information you will determine the absolute magnitude of the star at peak brightness.

PROCEDURE

1. Examine the accompanying data table (30-I). Notice that for each observation date, column three shows the elapsed time since the start of the supernova event on February 23, 1987.

2. Use either a metric ruler or vernier caliper to measure the supernova image diameter (to the nearest 0.1 mm) on each photograph at least twice and average your results. Record this average in your data table. Note that the brightest images of the supernova are surrounded by a halo. This halo is not connected to the star in any way and is only an artifact of the photographic process. By carefully examining each image, you should be able to distinguish the true image size from the surrounding halo.

3. On the graph paper provided, plot each of the SN1987 A image diameters versus the number of days since the supernova exploded. Draw a smooth curve through the plotted points. The date of maximum brightness may not have been recorded because of the timing of the photographs. Use your smoothed curve to estimate the date peak brightness occurred.

4. The usual way to determine the photographic magnitude of a star is to locate a number of stars of known magnitudes in the photographic field and measure their image diameters. The magnitude of a star can be estimated by comparing its diameter to the image diameters of stars of known magnitude. For the SN1987 A photographs provided here, the field does not contain any star as bright as the supernova at its peak brightness.

Figure 30-1 is provided to assist in estimating the peak magnitude of SN1987 A. It shows the image sizes that stars of known magnitude would have if they were photographed with the same system used to photograph the supernova.

5. Use your graph to determine the supernova's image diameter at peak brightness. Measure the diameters of the representative dots above and find the one that is closest in size to the supernova image at peak brightness. Estimate the apparent magnitude (m) of the supernova at its brightest.

6. Use the formula

$$M = m + 5 - 5 \log r$$

and your estimated peak apparent magnitude (m), calculate the absolute magnitude (M) of SN1987 A. Use the known distance to the LMC, 50,000 pc, as "r."

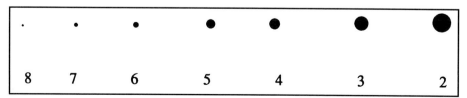

Figure 30-1. Reference Stellar Images. (The number below each image represents the magnitude of that image.)

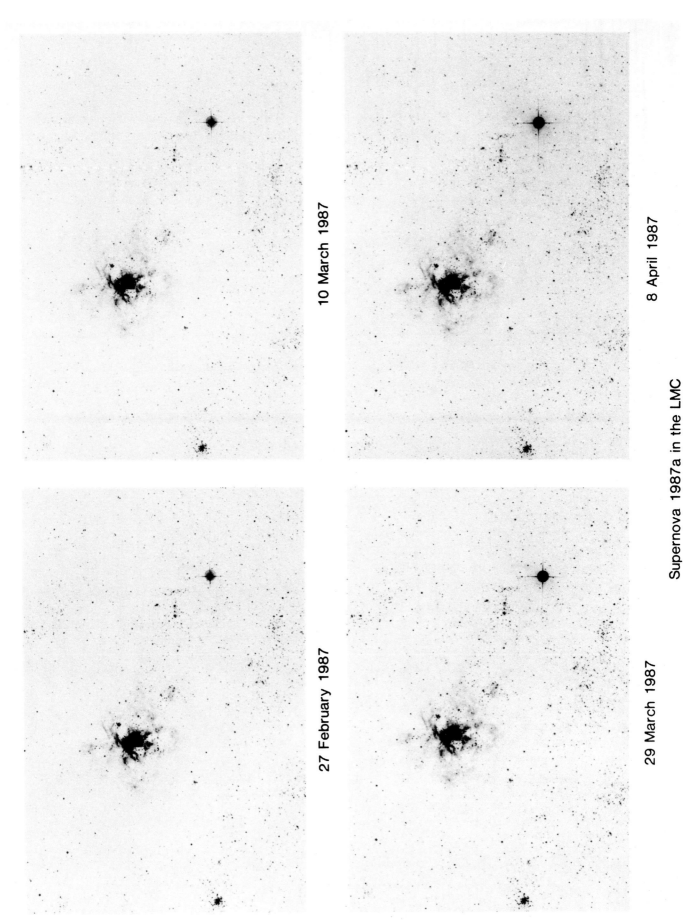

10 March 1987

8 April 1987

27 February 1987

29 March 1987

Supernova 1987a in the LMC

Figure 30-2. Supernova 1987A in the LMC.
(Copyright © Royal Observatory, Edinburgh, used with permission.)

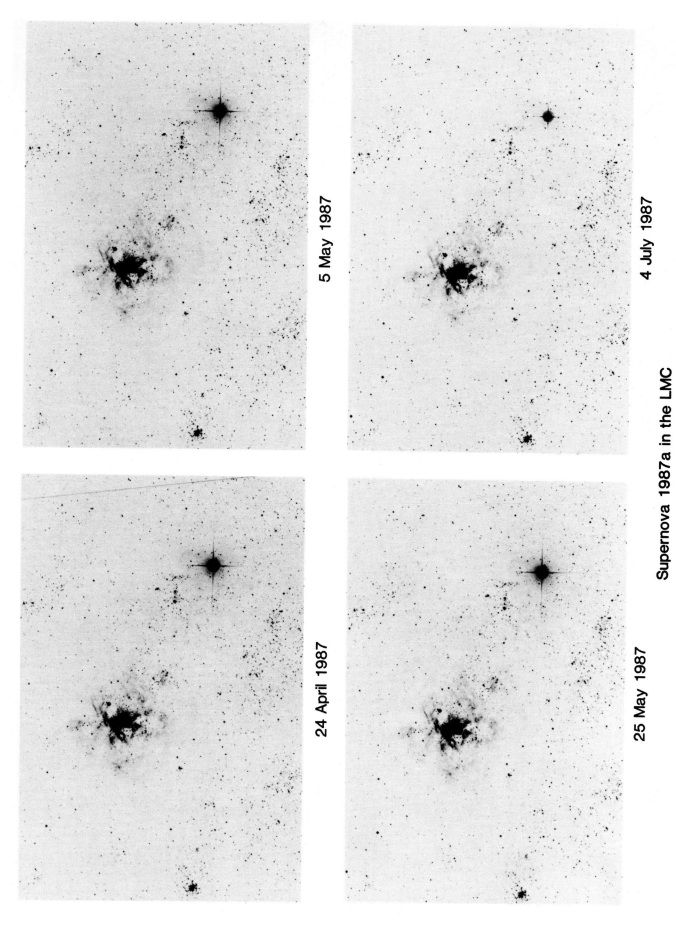

5 May 1987

4 July 1987

24 April 1987

25 May 1987

Supernova 1987a in the LMC

Figure 30-3. Supernova 1987A in the LMC.
(Copyright © Royal Observatory, Edinburgh, used with permission.)

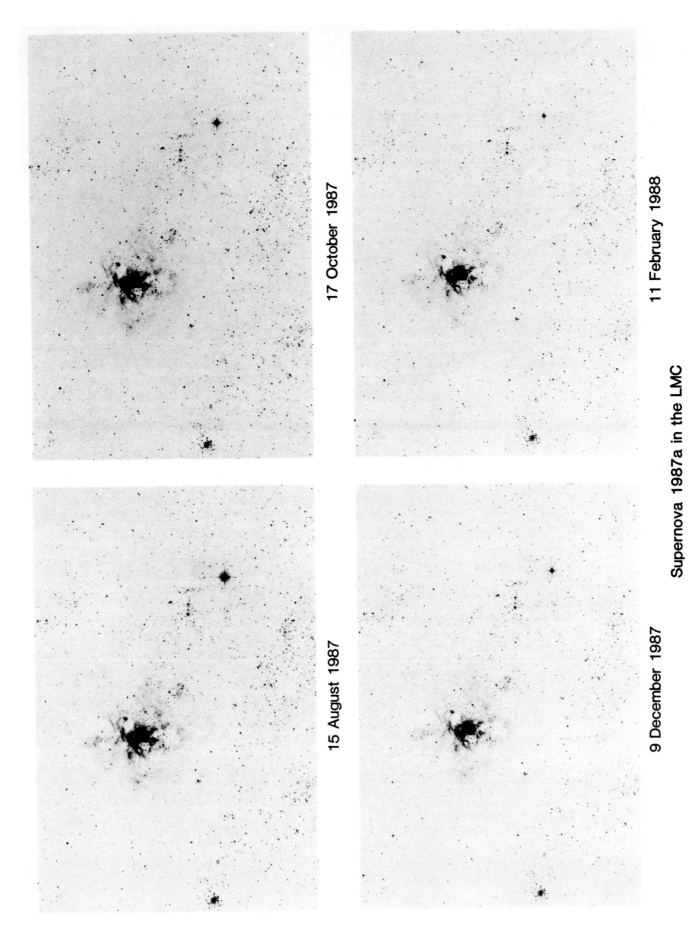

17 October 1987

11 February 1988

15 August 1987

9 December 1987

Supernova 1987a in the LMC

Figure 30-4. Supernova 1987A in the LMC.
(Copyright © Royal Observatory, Edinburgh, used with permission.)

30. Supernova 1987 A—Data Sheet

Photograph	Date	Days Since Event	Image Diameter (mm)		
			1st Measure	2nd Measure	Average
a	February 27, 1987	4			
b	March 10	15			
c	March 29	34			
d	April 8	44			
e	April 24	60			
f	May 5	90			
g	May 25	85			
h	July 4	131			
i	August 15	173			
j.	October 17	236			
k	December 9	289			
l	February 11, 1988	353			

1. After plotting the graph and drawing a smoothed line through the date points, determine the date of maximum brightness. What is that date? _____

2. Use your smoothed curve to arrive at an estimate of the image diameter of SN1987 A at the date of maximum brightness. Record the image size here _____ mm.

3. Using Figure 30-1, estimate the apparent magnitude (m) of SN1987 A at maximum brightness. At maximum brightness, SN1987 A had a magnitude of _____ .

4. Using the formula $M = m + 5 - 5 \log r$, calculate the absolute magnitude (M) of the supernova at peak brightness. Its absolute magnitude was _____ .

Show calculations and answers to discussion questions here.

DISCUSSION QUESTIONS

1. Based on your work, do you believe that SN1987 A was a Type I or Type II supernova?

2. What are some possible sources of error in this experiment?

30. Supernova 1987 A

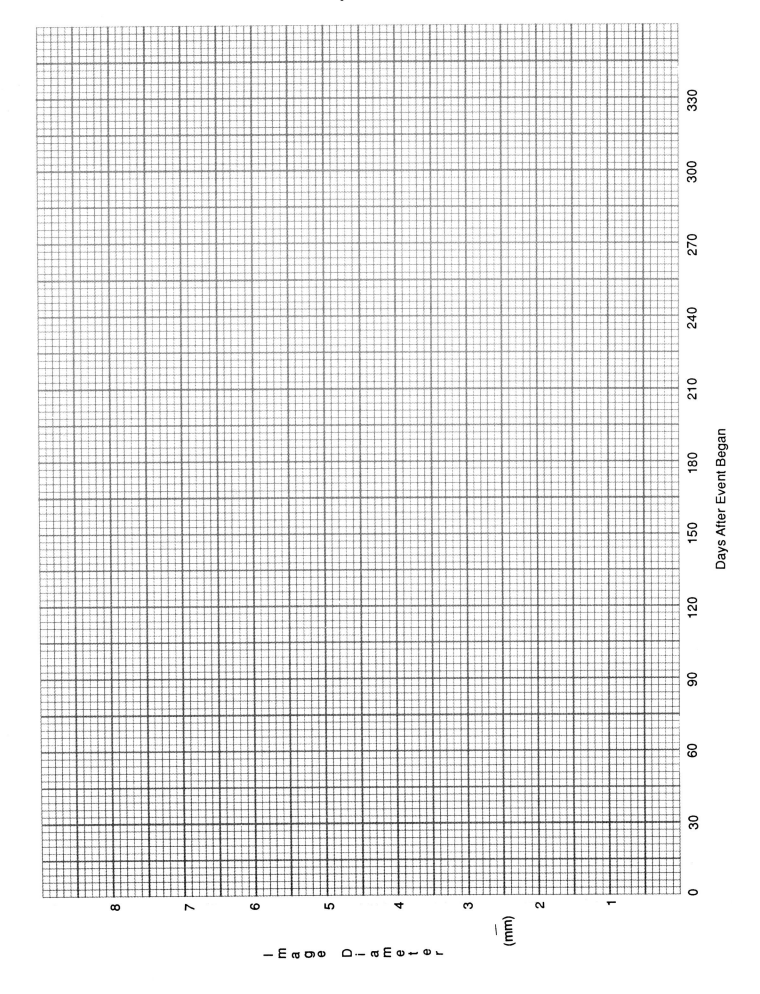

Days After Event Began

Image Diameter (mm)

31   Galactic Distances and Hubble's Law

PURPOSE AND PROCESSES

The purpose of this exercise is to study galactic recession and Hubble's velocity-distance law. The processes stressed in this exercise include:

Designing Experiments
Using Numbers
Interpreting Data
Identifying Variables
Formulation of Models
Prediction

INTRODUCTION

The precise determination of the distances to galaxies is not easy. Distances to some of the nearest galaxies are uncertain by ten percent, while the distances to most visible clusters of galaxies are uncertain by a factor of at least two.

The methods of galactic distance determination are indirect, and we must rely largely on properties of the brighter stars contained in galaxies. We assume that recognizable types of objects are similar in nearby and in distant galaxies of the same type. The period-luminosity relationship of cepheid variables can be used if a cepheid can be resolved in the galaxy. However, the brightest cepheids have magnitudes of about –5 at maximum and since we are limited to an apparent magnitude of 23 with the 200-inch telescope, we can use cepheids to a distance of 4 megaparsecs, including only galaxies in and very close to our local group. We can use the maximum magnitude of a nova or supernova to estimate galactic distances, but such events occur too infrequently to be used for a very systematic or extensive study. Some galaxies contain emission nebulae or H II regions that are brighter than the brightest stars, but they indicate only rough distances because it is not certain how luminous such nebulae can be. Likewise, estimates have been made by observing the brightest stars or globular clusters in a galaxy and assuming they are of comparable magnitude to those in nearby galaxies of a similar type. However, these values are also rather uncertain. Some methods of distance determination are summarized in Table 31-I.

Table 31-I
Distance Indicators for Galaxies

Object	M	Limiting Distance
Population II Red Giants	–2	1 Mpc
Brightest Cepheids	–5	4
Globular Clusters (Average)	–7	10
Globular Clusters (Brightest)	–9	25
Novae (Average)	–7	10
Novae (Brightest Common)	–9	25
Brightest Blue Stars	–9	25
Brightest H II Regions	–12	100
Super Novae, Type I	–18	1600

In 1929 Edwin Hubble found a correlation between the red shift or velocity of recession of galaxies and their distance, such that velocities are directly proportional to distances. Hubble used a brightest star criterion for his distance values. His law can be expressed as follows:

$$V = H \, r$$

where V = velocity of recession (km/sec)
    r = distance (megaparsecs)
    H = Hubble constant.

Hubble's **original** value of H was 550 km/sec per megaparsec. This says that a galaxy or cluster of galaxies moves away from us at a velocity of 550 km/sec for each million parsecs of its distance. This constant has been redetermined several times, each time "expanding" the size of the universe by several times its previously believed value.

PROCEDURE

1. Table 31-II contains a list of clusters of galaxies and their distances. Plot these data and determine a value for the Hubble constant. Note that two independently determined distances are given for several clusters. Estimate your uncertainty in H resulting from the scatter of the data.

Table 31-II
Approximate Distances and Velocities
of Clusters of Galaxies

	Distance (Mpc)	Velocity (Km/Sec)
Pegasus I	38	3,810
	46	3,860
Perseus	53	5,430
	80	4,960
Coma	58	6,657
	120	7,200
Hercules	105	10,400
Pegasus II	145	12,800
Gemini	228	23,400
	268	22,400
Leo	200	19,200
Ursa Major II	380	40,400
	450	40,000

2. Figure 31-2 shows spectra of five clusters of galaxies. Determine their red shifts and velocities of recession. A schematic of the comparison spectra identifying the various lines is given in Figure 31-1. Remember that you must:

(a) Measure the dispersion of the spectra.

(b) Measure the red shift in mm and convert to angstroms.

(c) Calculate the velocity using the Doppler equation:

$$v = c \, \frac{\Delta\lambda}{\lambda}$$

where v = velocity
    c = speed of light = 3×10^5 km/sec
    $\Delta\lambda$ = shift in wavelength
    λ = unshifted wavelength
      3968 Å
      = Ca II H and K
      3933 Å lines.

A review of Exercise 18 "Evidence of the Earth's Revolution" may be helpful as it uses the Doppler effect.

3. Determine the distance to each of the clusters given. Estimate your probable uncertainty for each.

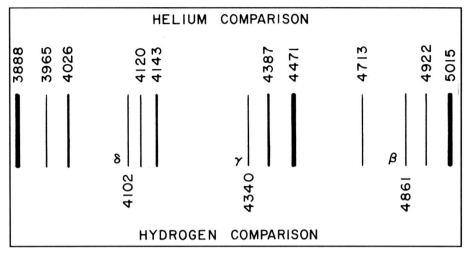

Figure 31-1. Line Identification for Comparison Spectra.

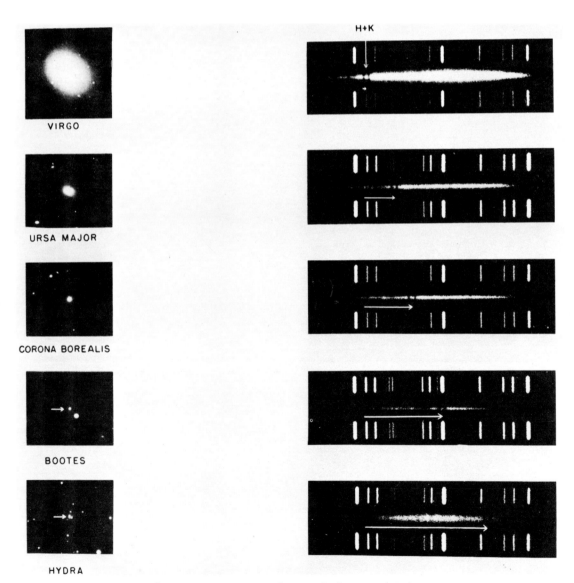

Figure 31-2. Spectra of Several Clusters of Galaxies.
(Courtesy of the Hale Observatories)

31. Galactic Distances and Hubble's Law

Velocity (km/sec)

Distance (Mpc)

DISCUSSION QUESTIONS

    1. What assumptions are being made about red shift in using Hubble's Law?

    2. Should the effects of special relativity be considered in your results? Explain. How will this affect estimates of the size of the universe?

32

Galaxies in the Virgo Cluster

PURPOSE AND PROCESSES

The purpose of this exercise is to develop a system for classifying galaxies, to compare the numbers of different types of galaxies in the Virgo cluster, and to determine the linear sizes of some of the cluster members. The processes stressed in this exercise include:

Classifying
Using Numbers
Prediction

INTRODUCTION

The Virgo cluster of galaxies is the nearest of the large clusters and is probably one of the largest. It covers an area of about ten degrees by twelve degrees in the sky and is centered near the Virgo-Coma constellation border. Nearly 3000 members have been identified on high-resolution photographs, and several are bright enough to be listed in the Messier catalogue.

The advantages of studying a cluster of galaxies are similar to those in studying star clusters: The objects are all at about the same distance so relative luminosities and diameters can be compared for the members. As with stellar evolution, in studying galaxies sharing a common origin, we hope to gain clues about the processes that govern the formation and evolution of different types of galaxies. We can also see how our galaxy or local group of galaxies compares with others in space.

PROCEDURE

Figures 32-1 through 32-4 are reprinted from a Palomar Sky Survey plate showing the Virgo cluster of galaxies. Other photos or slides of galaxies will be available in the classroom.

1. CLASSIFICATION OF GALAXIES

Examine as many photographs of galaxies as you can and look for similarities and differences. Consider properties such as overall shape, prominence of the central bulge relative to the disc, and possible arm structure.

(a) Devise a classification system for different types of galaxies. Use major classes and sub-classes if they seem appropriate.

(b) Devise a system of names or numbers to identify different classes.

(c) Could any of your classes represent the objects in other classes seen from different angles? That is, could one class be the same objects as another class but seen from a different point of view? Examine your classes for possible overlap due to different spatial orientations of the objects and revise your system if necessary to represent fundamentally different shapes of galaxies.

Figure 32-1. Portion of a Palomar Sky Survey Plate of the Virgo Cluster of Galaxies.
(Mount Wilson and Las Campanas Observatories, Carnegie Institution of Washington)

Figure 32-2. Portion of a Palomar Sky Survey Plate of the Virgo Cluster of Galaxies.
(Mount Wilson and Las Campanas Observatories, Carnegie Institution of Washington)

Figure 32-3. Portion of a Palomar Sky Survey Plate of the Virgo Cluster of Galaxies.
(Mount Wilson and Las Campanas Observatories, Carnegie Institution of Washington)

Figure 32-4. Portion of a Palomar Sky Survey Plate of the Virgo Cluster of Galaxies. (Mount Wilson and las Campanas Observatories, Carnegie Institution of Washington)

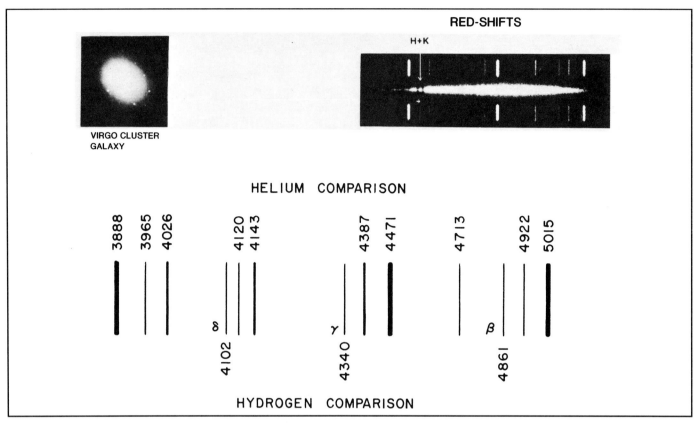

Figure 32-5. Spectrum of Virgo Cluster Galaxy and Hydrogen Comparison Spectrum. (Mount Wilson and Las Campanas Observatories, Carnegie Institution of Washington)

2. RELATIVE NUMBERS OF DIFFERENT TYPES OF GALAXIES

(a) A system is needed for describing general plate locations to avoid counting galaxies more than once. One method is similar to that used for archeological field work: Divide the plate into a grid and label or number the squares.

(b) Use your classification system to count and record the number of galaxies of each general class for the elements of your grid system. A hand magnifier may be helpful in distinguishing stars from giant galaxies. Galaxies have "fuzzy" edges while stars have clearer, more well-defined images.

(c) Calculate the proportion or percent of each major type of galaxy in the cluster.

3. DISTANCE TO THE CLUSTER

Figure 32-5 shows the spectrum of one galaxy in the Virgo cluster with the major comparison lines of hydrogen and helium identified.

(a) Determine the dispersion or plate scale for the galaxy spectrum in Å/mm. (You might want to refer to Exercise 18 "Evidence of the Earth's Revolution" for more detailed information on measuring Doppler shifts.)

(b) Measure the displacement of the calcium H and K lines (indicated by the arrow) in millimeters and convert this shift to Angstroms.

(c) Calculate the Doppler velocity for the spectrum using the relation

$$v = c \frac{\Delta\lambda}{\lambda}$$

where v = velocity
c = speed of light = 3×10^5 km/sec
$\Delta\lambda$ = shift in wavelength in Å
λ = unshifted or rest wavelength
= 3968 Å   Ca II H
= 3933 Å   and K lines.

(d) Calculate the distance to the Virgo cluster of galaxies using Hubble's law. Recall that

$$r = \frac{V}{H}$$

where r = distance in megaparsecs
V = the recessional velocity in km/sec
H = Hubble's constant = 100 km/sec/Mpc.

4. SIZE OF THE GALAXIES IN THE CLUSTER

It would be interesting to compare the linear sizes of some of the Virgo cluster galaxies to that of our own galaxy. Therefore it is necessary to determine the angular size of the galaxies. Remember that

$$s = r\theta$$

where s = linear diameter
r = distance
θ = angular diameter in radians.

(a) Measure the diameters of several of the galaxies to at least the nearest tenth millimeter using a low power measuring microscope or an eyepiece and recticle.

(b) The plate scale is 67.1″ of arc per mm.

(c) Calculate the angular diameters of the galaxies measured above.

(d) Calculate the linear diameters of the galaxies in kpc and compare these values to an estimated 30 kpc diameter for the Milky Way galaxy.

──────────────■──────────────

DISCUSSION QUESTION

1. How would the ratios of different types of galaxies differ for a more distant cluster where only the brightest galaxies are seen?

33 | The Absolute Magnitude of a Quasar

INTRODUCTION

Stars produce a wide range of electromagnetic radiation. For example, the sun (a typical star) radiates ultraviolet, visible, infrared, and radio waves. However, if the sun were placed at stellar distances, we would not expect to be able to detect the comparatively feeble radio waves it produces. It was with great surprise, therefore, that two STELLAR radio sources were detected in 1960. These sources, unlike many other radio sources, could be positively associated with star-like images which had been previously photographed in the visible part of the spectrum. These sources were first named quasi-stellar radio sources, a name later shortened to *quasars*.

Quasar observations also were perplexing for a second reason: Their visible spectra showed emission lines which could not be identified with known chemical elements. In 1963 M. Schmidt solved part of the problem by recognizing that the mysterious emission lines could be identified with known lines such as those of hydrogen if they were red shifted by very large amounts. However, if quasars obey accepted physical laws, such large red shifts would indicate that they must be receding at great velocities. Further, if they conform to Hubble's Law, all quasars must be at extremely great distances from our galaxy.

If we assume a quasar obeys Hubble's Law we can find its distance from its red shift. We can then observe its apparent magnitude and calculate the absolute magnitude. Such calculations produce incredible results. The interpretation of quasar data is one of the most controversial subjects in astronomy today. If we use Hubble's Law to interpret the red shifts, quasars produce more energy than any known source in the universe. If we instead assume that quasars are at closer distances, the incredibly large red shifts are impossible to explain. In each case, accepted physical laws and fundamental astronomical assumptions are in question.

In this exercise we will investigate the properties of 3C273, one of the first two quasars discovered. We will assume that it follows Hubble's Law and that its red shift is indicative of its distance. Remember, however, that there are prominent astronomers who disagree with this procedure and its basic assumptions.

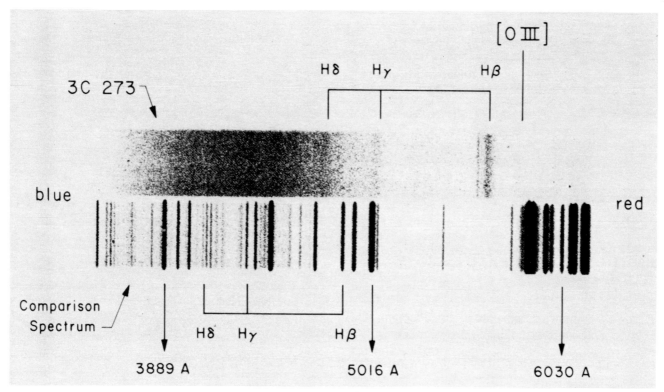

Figure 33-1. Spectrum of 3C273.
(Courtesy of the Hale Observatories)

PROCEDURE

Figure 33-1 shows the spectrum of quasar 3C273. The upper spectrum is that of the quasar and the lower spectrum is a comparison spectrum. The hydrogen Balmer lines (H_β, H_γ, and H_δ) are identified in the comparison spectrum and in their displaced positions in the quasar spectrum.

1. VELOCITY OF RECESSION
(a) Determine the dispersion or plate scale for the quasar spectrum in Å/mm. (You might want to refer to Exercise 18 "Evidence of the Earth's Revolution" for more detailed information on measuring Doppler shifts.)

(b) Measure the displacement of one of the hydrogen lines in the spectrum and convert this shift to Angstroms.

(c) Calculate the Doppler velocity for your spectral line. Remember that

$$v = c \, \frac{\Delta\lambda}{\lambda}$$

where v = velocity
c = speed of light = 3×10^5 km/sec

$\Delta\lambda$ = shift in wavelength in Å
λ = unshifted or rest wavelength in Å.

The rest wavelengths for the hydrogen Balmer lines are:

H_α = 6563 Å $\qquad$ H_γ = 4340 Å
H_β = 4861 Å $\qquad$ H_δ = 4102 Å

Repeat your measurements and calculations for the other hydrogen lines and average your results.

Note that for objects traveling at very high velocities this conventional Doppler equation must be replaced by the relativistic form. However, 3C273 is traveling slowly enough that the difference is not significant.

2. DISTANCE
Calculate the distance to 3C273 using Hubble's Law. Recall that

$$r = \frac{V}{H}$$

where r = the distance in megaparsecs
V = the recessional velocity in km/sec
H = Hubble's constant = 100 km/sec/Mpc.

Table 33-I

Comparison of Star Magnitudes

Star	Apparent Magnitude
a	12.5
b	13.1
c	12.1
d	12.6
e	13.2
f	12.8
g	13.7
h	14.0

Table 33-II

Absolute Magnitudes of Familiar Objects

Object	Absolute Magnitude
Sun	+4.8
Sirius	+1.5
Canopus	−5.0
Deneb	−7.0
Globular Cluster	−9
Irregular Galaxy	−18
Spiral Galaxy	−21
Elliptical Galaxy	−23

3. ABSOLUTE MAGNITUDE AND ENERGY OUTPUT

(a) Apparent Magnitude

Figure 33-2 shows a portion of the Palomar Sky Survey plate which contains 3C273 (marked with a black cross). This plate is a negative print on which bright objects appear dark and vice versa. Other stars in the field are marked with letters of the alphabet. Their apparent magnitudes are given in Table 33-I.

Estimate the apparent magnitude of 3C273, remembering that apparent photographic magnitudes are roughly inversely proportional to image size. (A graph of image diameter as a function of apparent magnitude may be useful).

(b) Absolute Magnitude

It is possible to calculate an object's absolute magnitude knowing its apparent magnitude and distance using the relationship

$$M = m + 5 - 5 \log r$$

Using your values for r (in parsecs) and your estimated apparent magnitude, calculate the absolute magnitude of 3C273.

(c) Energy Output

Table 33-II is a list of the absolute magnitudes of some familiar objects.

What conclusions can you make about the energy output of 3C273? Compare its brightness to that of several of the objects listed. (Remember that each magnitude difference corresponds to a brightness ratio of 2.5.)

4. VARIABILITY

Figure 33-3 shows the variation in apparent magnitude of 3C273 as measured from photographic plates since the late 1800s. The median value for these data corresponds with your measured apparent magnitude. Note the apparent regular variation in light output. It is generally accepted that the linear size of an object cannot exceed its period of variation. For example, an object with a period of 2 years can be no larger than 2 light years across.

(a) Estimate the period of variation of 3C273. What is its maximum size?

(b) Compare the absolute magnitude and size of 3C273 to those of other bright objects. Comment on this quasar's *energy density*, a quantity relating its energy output and size.

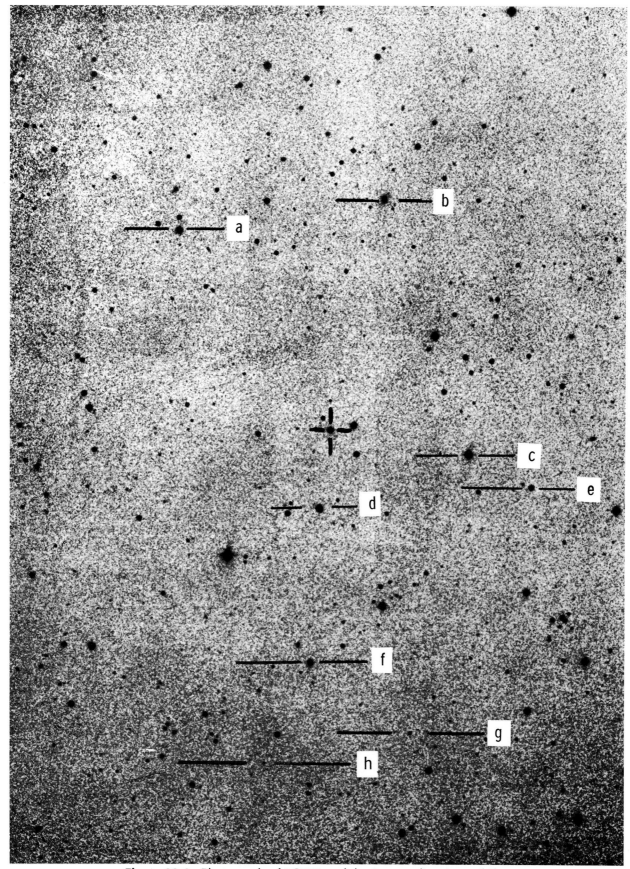

Figure 33-2. Photograph of 3C273 and the Surrounding Area of Sky.
(Courtesy of the Hale Observatories)

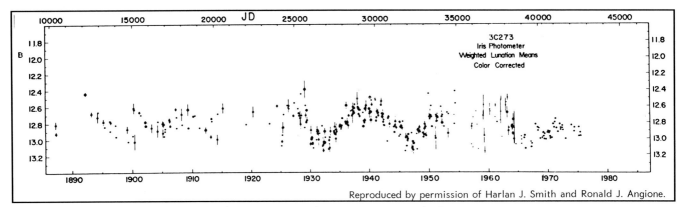

Figure 33-3. Brightness Variations of 3C273.

DISCUSSION QUESTIONS

1. List some possible sources of a quasar's red shift if it is NOT due to Hubble's Law and the expansion of the universe.

2. How do you think the quasar controversy will be resolved? What are some consequences of your choice on other parts of astronomy?

Appendices

Appendix 1. Suggestions for Writing Laboratory Reports

1. AT THE BEGINNING OF EACH REPORT GIVE
a. Title of the experiment.

b. Date.

c. Laboratory partner(s), if any.

2. PURPOSE
Summarize in a few words the purpose of the exercise, including any modifications as the exercise progressed.

3. REFERENCES
If any references other than those given in the laboratory guide are used, list them.

4. PROCEDURE
Summarize in a few lines the basic procedure used. If any substantial changes from the procedure suggested in the laboratory guide were made, note them.

5. DATA
a. Label all data.

b. Give units for all data.

c. List data in tabular form.

d. If possible, take several readings for each data point and use the average value.

6. GRAPH
a. Plot each graph to as large a scale as is practical.

b. Title each graph.

c. Label the quantities plotted on each axis and label each curve if you have more than one.

d. Give units on each axis.

e. Where appropriate, draw a smooth curve or line through the data points. DO NOT CONNECT THE POINTS IN A DOT-TO-DOT MANNER. The actual data points do not have infinite accuracy, and thus may not lie exactly on the proper curve. Draw a smooth curve such that positive and negative deviations are about equal and such that the curve matches the general trend of the data. This process averages the experimental fluctuations and the results deduced from the curve are usually more accurate than those deduced from individual measurements.

7. CALCULATIONS
a. Give calculations in a logical order down the page. Indicate the equation being used or the mathematical operation being done for each step.

b. Give units in each step of a calculation. Keeping close track of units may often help you to avoid errors.

c. If one method of calculation is repeated several times for different values, give a sample calculation and tabulate the results of the repeated calculations.

d. If a standard value is available for the quantity you have calculated, compare your experimental value to the standard and compute your percent of error. The percent of error is given by:

$$\frac{\text{standard-value} - \text{experimental value}}{\text{standard value}} \times 100\%$$

Note that this error in your derived value is not the "experimental error" resulting from uncertainty in measurements. Percent uncertainty or discrepancy between two values is calculated in a similar manner.

8. CONCLUSIONS
Give a brief statement of your conclusions and final results.

9. DISCUSSION QUESTIONS
Answer all questions asked in the lab guide, as well as any others your instructor presents, as concisely and completely as possible. THINK before you write.

Appendix 2. Alignment of an Equatorial Telescope

INTRODUCTION

The most commonly used type of telescope mounting is the equatorial mount, based on the Equatorial Coordinate System. This mount has a *polar axis* which is adjusted to the altitude and azimuth of the north celestial pole. The altitude of the axis, which must be in the plane of the meridian, is set equal to the observer's latitude. The *declination axis* is permanently affixed to the polar axis and requires adjustment only on its setting circles.

The advantage in using an equatorial mount is that as time passes, a star's apparent motion is parallel to the celestial equator. Thus, a simple clock drive can produce rotation about the polar axis at a rate equal and opposite to that of the earth's rotation, so that a star will remain in the field of view. If the polar axis of the mount is improperly aligned, the star will drift out of the field of view as the earth rotates.

PROCEDURE

1. ALIGNMENT OF THE PRIMARY MIRROR
(a) Remove the eyepiece.

(b) Look into the telescope (without the eyepiece) at the secondary mirror. You will see an image of the secondary mirror and its supports reflected from the primary mirror.

(c) Adjust the screws located on the back of the primary mirror until all of the supports of the secondary mirror appear to be of the same length when viewed as described above (Figure A2-1). The optical axis of the primary mirror will now be centered and parallel to the telescope tube.

2. BALANCING THE TELESCOPE
Put an eyepiece in the holder (the drive motor remains turned off for this and the remaining steps in the alignment). Balance the telescope tube by sliding it up and down in the tube brackets until the telescope stays in any direction pointed.

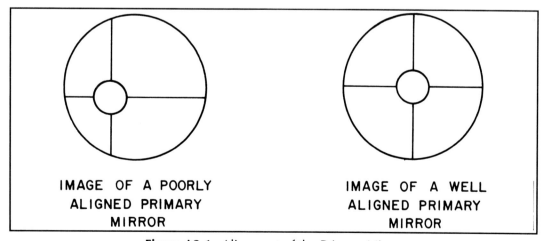

IMAGE OF A POORLY ALIGNED PRIMARY MIRROR

IMAGE OF A WELL ALIGNED PRIMARY MIRROR

Figure A2-1. Alignment of the Primary Mirror.

3. ALIGNMENT OF THE POLAR AXIS

(a) Altitude

The altitude of the polar axis must be set equal to the observer's latitude, and if the set screws are sufficiently tight, it should not need to be reset until the mount is moved. For visual observation it is usually sufficiently accurate to use Polaris for this setting; otherwise, the deviation of Polaris from the true pole should be taken into account.

   i. Point the polar axis toward Polaris and set the telescope so it is on top of the polar axis and accurately parallel to it. (The counterweight will be underneath the axis.)

   ii. Adjust the altitude of the polar axis until Polaris is accurately in the center of the field of view. Tighten the set screws securely.

(b) Azimuth

Unless a permanent mount is available this critical alignment needs to be done each time the telescope is used.

   i. Rotate the telescope mount until the line joining the top and bottom of the polar axis is approximately parallel to a north-south line.

   ii. Point the telescope toward Polaris with the declination axis parallel to the horizon and the counterweight toward the east. Rotate the telescope in its mount until Polaris is nearly on the vertical cross hair. Tighten the screws securing the telescope base into the mount.

   iii. Adjust the fine-adjust screws located on the telescope base until Polaris is exactly on the vertical cross hair. Make sure the declination axis remains parallel to the horizon.

   iv. Rotate the telescope so that the counterweight is on the west side of the mount, and again place the declination axis parallel to the horizon. Polaris should be near the center of the field of view; re-center it using the fine-adjust screws.

   v. Repeat the previous two steps with the counterweight on opposite sides of the mount each time, until Polaris is on the vertical cross hair in both positions. IT IS ESSENTIAL THAT THE DECLINATION AXIS BE PARALLEL TO THE HORIZON EACH TIME AN ADJUSTMENT IS MADE.

   vi. This is an approximate method, accurate enough for visual use of the telescope. If the telescope is to be used for astronomical photography, a more accurate alignment method (such as that described in *Norton's Star Atlas*) should be used. In addition, the fact that Polaris is nearly one degree from the north celestial pole must be accounted for in using the telescope for long time exposures.

4. ALIGNMENT OF THE FINDER TELESCOPE

It is necessary to adjust the screws holding the finder telescope so it and the main telescope point in the same direction. Find a distant identifiable object (a light or other object near the horizon works well) and center it in the field of the main telescope. Adjust the finder telescope until the object is accurately centered on the cross hairs.

It is important to use a distant object for this procedure so that the parallax effect is negligible.

5. ADJUSTMENT OF THE SETTING CIRCLES

Center the telescope on a bright star of known coordinates and adjust the declination and hour circles to the correct readings.

Appendix 3. Equipment Notes

Here are some informal notes on our experiences with equipment. The companies listed are not necessarily the best sources for any item, nor is this meant to be an exhaustive list. It is simply those items we are most familiar with. Prices given are those in effect during the spring of 1991 and may not be up-to-date.

1. VISUAL ASTRONOMY AND STAR COUNTS

a. Constellation Charts: Extra SC1 and SC2 charts are available from Sky Publishing Company at $0.30 and less in quantity orders.

b. Pathfinder or Star Locators: Available from Edmund Scientific Co. at $2.75 (cheaper in quantity). A plastic planisphere (Phillips Planisphere) is available from Optica Company at about $9.95. (Be sure to state your latitude when ordering.) Lawrence Hall of Science distributes *Sky Challenger Star Wheels*. These are regular star finders; in addition insertable wheels are provided for locating binocular objects and for showing native American constellations. They sell for about $6.95. The *Observer's Handbook* is available for $13.50 from the Royal Astronomical Society of Canada. Guy Ottewell's *Astronomical Calendar* is available from Furman University or Sky Publishing for $15.00. Astronomical wall calendars are available from AstroMedia, Sky Publishing, or Hansen Planetarium for $8.95 to $9.95.

c. Binoculars

d. Small telescope (if possible)

e. Cardboard mailing tube (2–4 inch diameter, about 18 inches long works well)

f. Table of random 5-digit numbers (optional)

The star counting part of the exercise can also be done in a planetarium using sections of paper towel tubes.

g. References listed in exercise

2. OBSERVING EXERCISE I: THE MOON

a. Telescope

b. Additional moon maps. Moon maps and descriptions are found in many handbooks, field guides, and annuals.

3. OBSERVING EXERCISE II: THE PLANETS

a. Telescope

b. *The Astronomical Almanac, Sky and Telescope, Astronomy,* or *Observer's Handbook* for planet positions.

c. Camera, film, and tripod (optional)

You might want to suggest planets to observe or photograph during the semester or quarter. Venus is excellent (if visible in the evening!) and try to suggest the other planets as they retrograde (near opposition).

d. Cross-staff or plastic strip for measuring angles, star chart (optional)

4. OBSERVING EXERCISE III: THE SUN

a. Telescope

b. Solar filter or cardboard ring and projection screen

c. Camera (optional)

Mylar material from camping "space blankets" can be used to make solar filters. It may require multiple layers. Carefully test the opacity before allowing student use.

5. OBSERVING EXERCISE IV: STARS, CLUSTERS, AND NEBULAE

a. Telescope

b. Erfle eyepiece, if possible

c. The Sky Challenger (see 1 of this Appendix) provides an excellent reference set of the most easily located objects.

6. ASTRONOMICAL PHOTOGRAPHY

a. Film: Tri-X, T-MAX, Plus-X, Ektachrome 200 or 400. These films are available from Eastman Kodak Company in bulk rolls. Tri-X film currently sells for about $34.00 (T-MAX for $36.00) for 100-foot rolls. Reusable film canisters are available from most photography shops at about $1.00, and daylight bulk film loaders can be obtained for $12.00 to $25.00. Loading your own film provides substantial savings, and 100-foot rolls last for a considerable period if refrigerated.

In addition, self-loaded 10-exposure rolls of film are often more convenient for astronomical work than longer commercial rolls. Some new color films include Kodak 2415 and Kodak Ektar 1000, Fujicolor 1600 and Fujicolor HG-400, and Konica SR-V3200 and Konica SR 400.

b. Camera: Any Polaroid with an electric eye works well. Some commonly used SLR cameras are Canon, Pentax, Nikon, and Olympus.

c. Barlow lenses: Available from most astronomical and optical supply houses. (Many are advertised in *Sky and Telescope* and *Astronomy.*)

d. Adaptors: Two basic adaptors are needed. You will need a "T-adaptor" to fit your particular camera. These are typically available from camera supply houses at about $16.95. You will also need a telescope adaptor to fit the diameter of your particular telescope eyepiece holder. These are available from Edmund for $16.95 for 1¼ inch, and $17.95 for the 0.965-inch eyepiece holders.

7. FIELD OF VIEW OF A TELESCOPE

a. Telescope (with drive if possible)

b. Slow-motion control (optional)

c. Stopwatch

d. List of stars at various declinations for groups of students to observe.

8. PLOTTING THE MOON'S ORBIT

a. Constellation chart (see 1 of this Appendix)

b. Cross-staff or plastic strip, about ¾ inch by 18 inches. Strips can be cut from nearly any flexible plastic material, but clear strips are of some advantage. Ribbed plastic matting (often used in chemistry laboratories) and plastic window coverings have been used satisfactorily and are very inexpensive. See Exercise 10 for calibration techniques.

c. Compass

9. STAR CHARTS AND CATALOGUES

a. Constellation Charts: See 1 of this Appendix.

b. The *Sky Gazer's Almanac* appears as an 11″ × 17″ color graphic in the January issue of *Sky and Telescope* magazine and as part of the *Current Guide to the Heavens* for $1.00 (less if ordered in quantity) and as a black and white wall poster for $4.95, all available from Sky Publishing.

c. Rotating star and planet locators (see 1 of this Appendix)

d. Various charts and catalogues listed in the exercise itself.

10. ANGLES AND PARALLAX

a. Plastic Strip; about ¾ inch by 18 inches. Strips can be cut from nearly any flexible plastic material, but clear strips are of some advantage. A ribbed plastic matting (often used in chemistry) has been used satisfactorily, is very inexpensive.

b. Graph Paper

11. KIRCHHOFF'S LAWS AND SPECTROSCOPY

a. Spectroscopes: Holographic grating material for making your own is available in bulk ($5.00 for a 5″ × 9″ sheet and $25.00 for a 5″ × 6′ roll) and in 35-mm glass slide mounts ($30.00 for 10) from Learning Technologies. Cardboard spectroscopes with regular plastic gratings are available from various science supply houses for $30–$40 for class sets of 15.

b. Gas Discharge Tubes, Holders, and Power Supplies: We suggest tubes of as many of the following gasses as possible: H, He, Na, Hg, Ni, Air, O, H_2O, Xe, Kr. If possible, it is good to have spares, and you will want extras of several of them to use as "unknowns." Tubes are available from various science supply houses for $18–$26 and corresponding power supply for $140–$155. Fancy screw-in tubes are available from CENCO at $140–$220 for the lamps and $500 for power supply and lamp case.

c. This setup can be used if gas discharge tubes are not available:

Bunsen Burners

Platinum Test Wires

Salts such as CaO, SrBr, KBr

Dilute HC1 (for cleaning test wires)

12. IMAGE SIZE—FOCAL LENGTH RELATIONSHIP

An optician's millimeter ruler and a vernier caliper are available from Edmund Scientific for $5.75 and $10.50, respectively.

13. LENGTH OF THE SIDEREAL DAY

a. Camera, film, stopwatch, and tripod (optional)

b. Polar coordinate tracing graph paper (optional) or a compass

14. DETERMINING THE MASS OF THE MOON

If a calculator-plotter or computer is available it can help relieve the tedium of plotting the many data points.

15. LUNAR FEATURES AND MOUNTAIN HEIGHTS

a. Camera, film, telescope to photograph moon (optional). It might be of value to plan a sequence of labs to photograph the moon in one session, develop and print the photographs in the next session, and to do this lab as follow-up.

b. Commercial lunar photos (optional): Black-and-white wall-poster size photos of the first and third quarter moon are available from various science supply houses for $11.40–$16.25 per set.

c. Ruler and/or meter stick

d. *The American Ephemeris and Nautical Almanac* or *The Astronomical Almanac*

16. THE MOON'S SIDEREAL PERIOD

a. Camera, film and tripod (if moon passes close enough to a bright star or planet to obtain your own photographs)

b. Tracing paper

c. Compass

17. THE MOON'S GEOLOGIC HISTORY

a. Scissors

b. Transparent tape

18. EVIDENCE OF THE EARTH'S REVOLUTION

Millimeter ruler or magnifier with reticle: A flat plastic ruler works best. Various magnifiers and reticles are available from Edmund starting at $37.00

19. COLLECTING MICROMETEORITES

a. Microscopes with 25× to 30× magnification

b. Microscope slides

c. Donut magnets

d. Light cord or string

e. Petroleum jelly

f. Petri dishes or other flat glass containers

g. Bamboo skewers

h. Small screened sieves.

20. HEIGHT OF A METEOR

a. Camera, film and tripod

b. Stop watches or timing devices

c. Star charts

d. Millimeter ruler

21. MERCURY'S ROTATIONAL PERIOD

Millimeter ruler

22. MEASURING THE DIAMETER OF PLUTO AND CHARON

No special equipment needed.

23. DETERMINING THE VELOCITY OF A COMET

Millimeter ruler

24. SOLAR ROTATION

a. Tracing paper

b. Ruler (mm) or magnifier and reticle (see 17 of this appendix)

25. PROPER MOTION OF A STAR

a. This lab uses a piece of translucent graph paper found in Appendix 4.

b. Graph paper (mm)

c. Compass

d. Ruler

26. SPECTRAL CLASSIFICATION

No special equipment needed.

27. A COLOR-MAGNITUDE DIAGRAM OF THE PLEIADES

Camera (optional)

28. DISTANCE TO THE PLEIADES

Ruler and compass (optional)

29. GALACTIC CLUSTERS AND HR DIAGRAMS

Graph paper
Note: This lab uses a transparency found in Appendix 4.

30. SUPERNOVA 1987 A

Millimeter ruler or (preferred) vernier caliper

31. GALACTIC DISTANCES AND HUBBLE'S LAW

Ruler (mm) or magnifier and reticle (see 17 of this appendix)

32. GALAXIES IN THE VIRGO CLUSTER

a. Palomar Sky Survey Print (+12°, 12^{h}24^m). This is one print of a set of six available from the California Institute of Technology. A set of six different prints including the one referenced sells for $31.65 plus $3.50 for shipping. This set also includes other prints useful in the laboratory.

b. Ruler (mm) or magnifier with reticle (see 17 of this appendix)

33. ABSOLUTE MAGNITUDE OF A QUASAR
a. Ruler (mm) or magnifier and reticle (see 17 of this appendix)

b. Graph Paper

ADDRESSES OF SUPPLIERS REFERRED TO IN THIS APPENDIX
Astronomical Education Materials
AstroMedia Corporation
21027 Crossroads Circle
Box 1612
Waukesha, WI 53187-1612
(414) 796–8776

Astronomical Society of the Pacific
390 Ashton Ave.
San Francisco, CA 94112
(415) 337–1100

California Institute of Technology
Bookstore 1-51
Pasadena, CA 91125
(818) 356–6161

Discovery Corner
Lawrence Hall of Science
University of California Berkeley
Berkeley, CA 94720
(415) 642–1016

Hansen Planetarium
1098 South 200 West
Salt Lake City, UT 84101-9917
(801) 538–2104

MMI Corp.
2303 N. Charles
Baltimore, MD 21218
(301) 366–1222

Optica Co.
4100 MacArthur Blvd.
Oakland, CA 94619
(415) 530–1234

Sky Publishing Corporation
P.O. Box 9111
Belmont, MA 02178-9111
(617) 864–7360

Willmann-Bell, Inc.
P.O. Box 35025
Richmond, VA 23235
(804) 320–7016

Government Printing Office
Washington, DC 20402
(202) 783–3238

General Science Supplies
Central Scientific Co.
11222 Melrose Ave.
Franklin Park, IL 60131-1364
(800) 262–3626

Edmund Scientific Co.
101 E. Gloucester Pike
Berrington, NJ 08007-1380
(609) 573–6270

Fisher EMD
4901 W. LeMoyne St.
Chicago, IL 60651
(800) 621–4769

Frey Scientific
905 Hickory Ln.
P.O. Box 8101
Mansfield, OH 44901-8101
(800) 225–3739

Sargent-Welch
7350 N. Linder Ave.
Skokie, IL 60076
(800) 727–4368

Science Kit & Boreal Laboratories
777 E. Park Dr
Tonawanda, NY 14150-6784
(800) 828–7777

Optical Companies
Astro-Physics
7470 Forrest Hills Rd.
Loves Park, IL 61111
(815) 282–1513

Celestron International
2835 Columbia St.
Torrence, CA 90503
(800) 421–1526

Coulter Optical, Inc.
P.O. Box K
Idyllwild, CA 92349-1107
(714) 659–4621

Meade Instruments Corporation
1675 Toronto Way
Costa Mesa, CA 92626
(714) 556–2291

Parks Optical
270 Easy St.
Simi Valley, CA 93065
(805) 522–6722

Questar
P.O. Box 59
Dept. 118
New Hope, PA 18938
(215) 862–5277

Tele Vue Optics, Inc.
20 Dexter Plaza
Pearl River, NY 10965
(914) 735–4044

Astronomical and Optical Supplies
Adorama
42 West 18th St.
New York, NY 10011
(212) 741–0052

Astronomics
2401 Tee Circle
Suites 105/106
Norman, OK 73069
(405) 364–0858

Jim's Mobile Inc.
1960 County Rd.
Evergreen, CO 80439
(303) 277–0304

Lumicon
2111 Research Dr. #5
Livermore, CA 94550
(415) 447–9570

Orion Telescope Center
2450 17th Ave.
P.O. Box 1158-S
Santa Cruz, CA 95061
(800) 447–1001

Roger W. Tuthill, Inc.
Box 1086
Mountainside, NJ 07092
(800) 223–1063

Appendix 4. SC1–SC2 Charts, and Starfinder Parts Translucent Graph Paper

SC001 CONSTELLATION CHART
EQUATORIAL RELIGION — EPOCH 2000

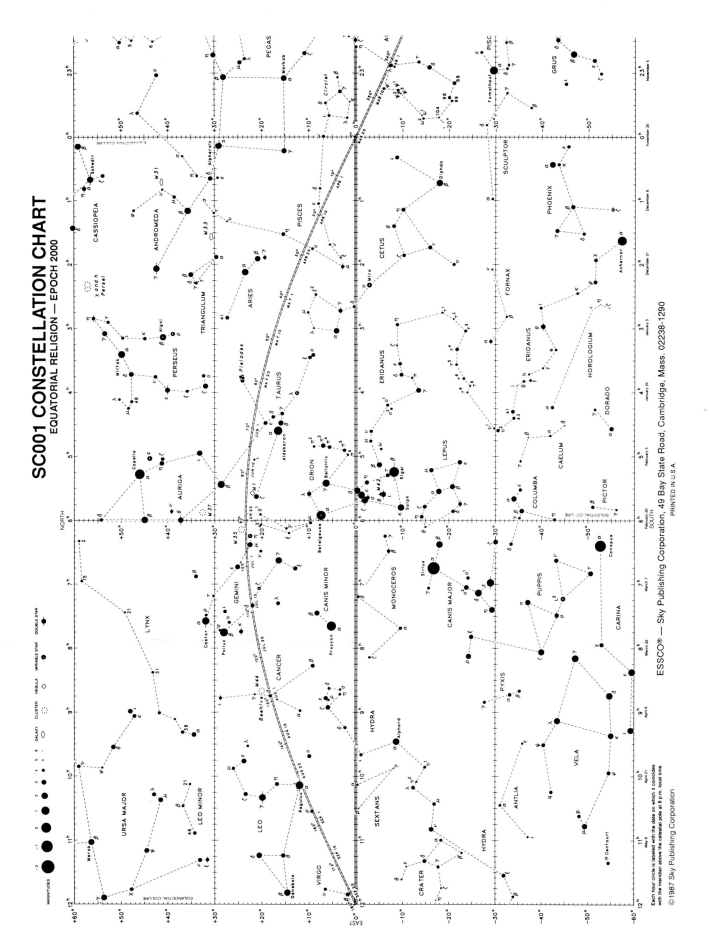

ESSCO® — Sky Publishing Corporation, 49 Bay State Road, Cambridge, Mass. 02238-1290

PRINTED IN U.S.A.

Each hour circle is labeled with the date on which it coincides with the meridian above the celestial pole at 8 p.m. local time.

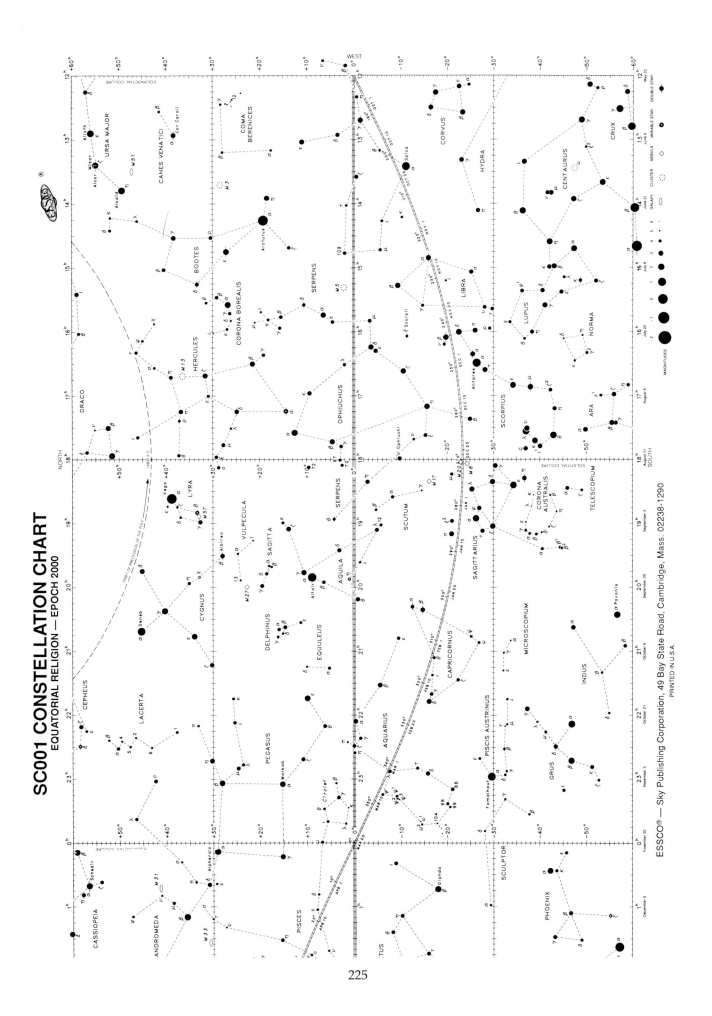

SC001 CONSTELLATION CHART
EQUATORIAL RELIGION — EPOCH 2000

ESSCO® — Sky Publishing Corporation, 49 Bay State Road, Cambridge, Mass. 02238-1290

PRINTED IN U.S.A.

225

SC002 CONSTELLATION CHART

NORTH CIRCUMPOLAR REGION — EPOCH 2000

FROM 30° N TO 90° N

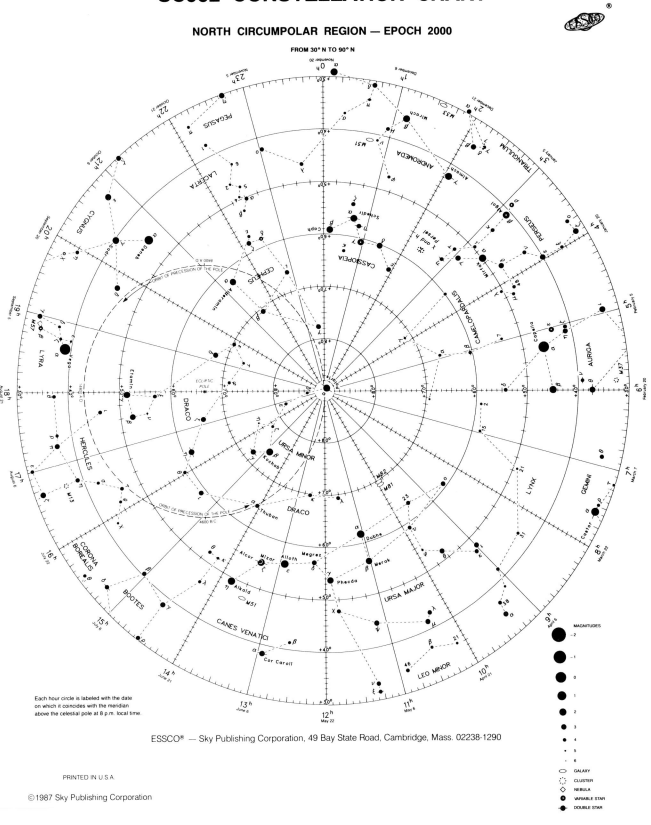

Each hour circle is labeled with the date
on which it coincides with the meridian
above the celestial pole at 8 p.m. local time.

MAGNITUDES

- −2
- −1
- 0
- 1
- 2
- 3
- 4
- 5
- 6

GALAXY
CLUSTER
NEBULA
VARIABLE STAR
DOUBLE STAR

ESSCO® — Sky Publishing Corporation, 49 Bay State Road, Cambridge, Mass. 02238-1290

PRINTED IN U.S.A.

227

28. Distance to the Pleiades

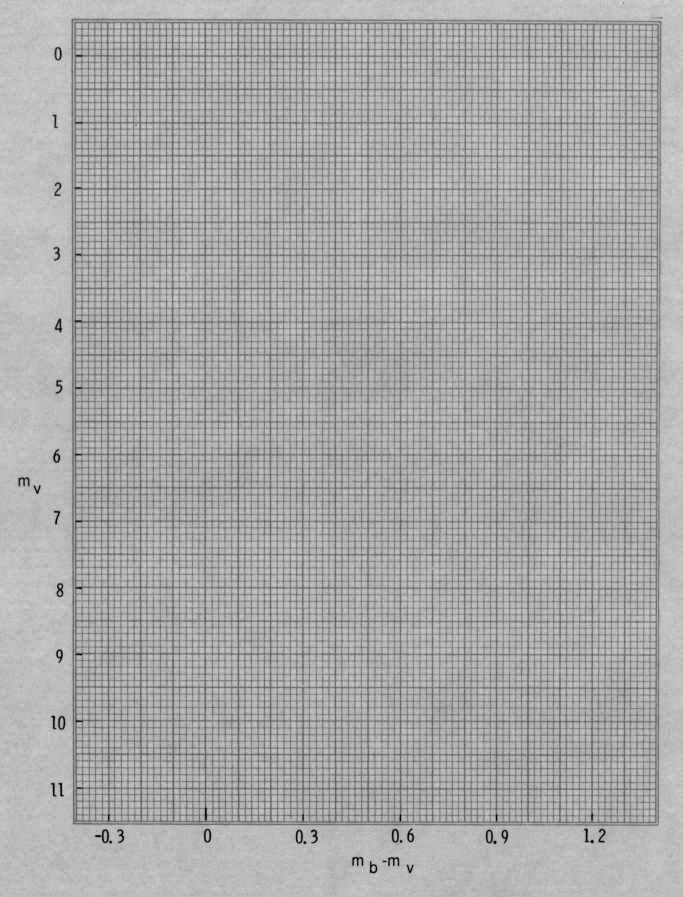

28. Distance to the Pleiades

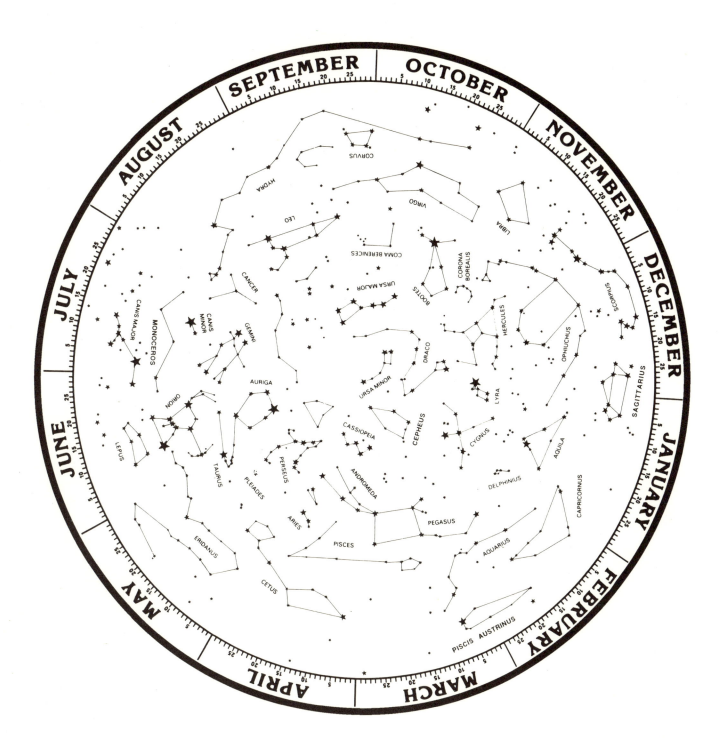

Do It Yourself Star Finder

To use, turn the star map until the current date lines up with the correct time. Hold the chart in front of you with the "NORTH" corner down if you are facing north, or the "SOUTH" corner down if you are facing south.

CUT OUT THIS AREA

6 am
5 am
4 am
3 am
2 am
1 am
12 midnight
11 pm
10 pm
9 pm
8 pm
7 pm

EAST
NORTH
WEST
SOUTH

CUT OUT THIS AREA

PASTE THIS FLAP BACK

FOLD ALONG THIS LINE

FOLD ALONG THIS LINE

CUT OUT THIS AREA

The *Do It Yourself Star Finder* is an adaptation of the *Sky Challenger* which was designed by Budd Wentz and adapted for classroom use by Edna DeVore of the Lawrence Hall of Science, University of California at Berkeley, Berkeley, CA 94720. The *Sky Challenger* was originally developed under National Science Foundation Grant #SED 77-18818. ©1978 Regents of the University of California.